Falling for Science

Falling for Science

Objects in Mind

edited and with an introduction
by Sherry Turkle

The MIT Press   Cambridge, Massachusetts   London, England

For information about special quantity discounts, please email special_sales@mitpress.mit.edu.

This book was set in Bookman Old Style, ITC Bookman, and Stymie by Graphic Composition, Inc., Bogart, Georgia.

Printed and bound in the United States of America.

Library of Congress Cataloging-in-Publication Data

Falling for science : objects in mind / edited and with an introduction
    by Sherry Turkle.
        p. cm.
    Includes bibliographical references and index.
        ISBN 978-0-262-20172-8 (hardcover : alk. paper) 1. Science—
    Study and teaching—United States. 2. Engineering—Study and
    teaching—United States. 3. Technology—Study and teaching—
    United States.
    I. Turkle, Sherry.
    Q181.3.F35 2008
    500—dc22

                                                        2007020843

10 9 8 7 6 5 4 3 2 1

To Seymour Papert

Contents

x Acknowledgments

3 Sherry Turkle | *Introduction: Falling for Science*

PART I
MIT Students and Their Objects (1979–2007)
What Makes A Scientist?

WHAT WE SEE
42 Steven Schwartz | *Maps* (1982)
45 Thomas P. Hermitt | *Prisms* (1984)
47 Jennifer Beaudin | *Walls* (2002)
49 Andrew Sempere | *Holga Camera* (2004)

WHAT WE SENSE
54 Stephen Intille | *Sand Castles* (1992)
58 Joanna Berzowska | *The Body* (1996)
60 Matthew Grenby | *Bubbles* (1996)
63 Jonah Peretti | *Clay* (1999)
66 Diane Willow | *Mud* (1999)

WHAT WE MODEL
70 Lauren Seeley Aguirre | *Gumby* (1985)
73 Michael Murtaugh | *Easy-Bake Oven* (1990)
76 Emmanuel Marcovitch | *Kitchen Clock* (1996)
80 Christine Alvarado | *My Little Pony* (1999)
82 Cameron Marlow | *Fly Rod* (1999)
84 James Patten | *Wood Stove* (1999)
88 Daniel Kornhauser | *Shirts* (2000)
92 Mara E. Vatz | *Vacuum Tubes* (2003)
96 Selby Cull | *Chocolate Meringue* (2006)

WHAT WE PLAY

102 Walter Novash | *Cords* (1982)

105 Jonah Benton | *Dice* (1990)

108 Erica Carmel | *Egg Basket* (1992)

111 Eric Choi | *Keys* (1992)

114 Janet Licini Connors | *Cardboard Boxes* (1992)

118 Gil Weinberg | *Music Box* (1997)

120 Kwan Hong Lee | *Marbles* (1999)

123 Robbin Chapman | *Jacks* (2000)

126 Douglas Kiang | *Pachinko Machine* (2000)

WHAT WE BUILD

132 Chuck Esserman | *Bikes* (1979)

134 Kwatsi Alibaruho | *Erector Set* (1990)

137 Austina (Vainius) De Bonte | *Straws* (1996)

142 Timothy Bickmore | *Lasers* (1998)

147 Justin Marble | *LEGO Particles* (1982)

150 Sandie Eltringham | *LEGO Metrics* (1990)

153 Alan Liu | *LEGO People* (1990)

156 Andrew Chu | *LEGO Laws* (1992)

159 Scott Brave | *LEGO Planning* (1996)

162 Dana Spiegel | *LEGO Replicas* (1996)

164 Joseph "Jofish" Kaye | *LEGO Categories* (1998)

WHAT WE SORT

168 Todd Strauss | *Wallpaper* (1982)

170 Britt Nesheim | *Toy Mailbox* (1989)

173 Joseph Calzaretta | *Stop Signs* (1992)

175 Brian Tivol | *Cards* (1996)

WHAT WE PROGRAM

180 Fred Martin | *BASIC Manual* (1989)

183 David (Duis) Story | *Apple II* (1990)

189 Ji Yoo | *Atari 2600* (1994)

193 Chris Dodge | *TRS-80* (1996)

199 Nelson Minar | *BASIC* (1996)

203 Steve Niemczyk | *Atari 800* (1996)

206 Rachel Elkin Lebwohl | *Apple II* (1998)

208 Anthony Townsend | *Modem* (1999)

212 Antoinne Machal-Cajigas | *TurboGrafx 16* (2007)

PART II
Mentors and Their Objects
What Made a Scientist?

220 Susan Hockfield | *What We See: Microscope*

228 Rosalind Picard | *What We Sense: Purple Haze*

236 Moshe Safdie | *What We Model: Steps*

242 Sarah Kuhn | *What We Play: Blocks*

248 Donald Norman | *What We Build: Radio*

252 Donald Ingber | *What We Sort: Venus Paradise Coloring Set*

262 Alan Kay | *What We Program: Vacuums*

268 Seymour Papert | *Objects in Mind: Gears*

273 Sherry Turkle | *Epilogue: What Inspires?*

284 Notes

291 Bibliography and Suggested Readings

305 Illustration Credits

306 Index

Acknowledgments

For more than twenty-five years I have been asking MIT students to share with me early moments of scientific curiosity. Their answers changed my experience of science and deepened my commitment to understanding the role of objects in the creative process. Here I share some of their stories so that others may benefit as well. I am grateful to all my students and thankful that all those I asked for permission to publish their essays said yes. I also thank the mentor scientists who, when I explained what this book was about, understood at once and wrote with such intellectual and emotional generosity.

At MIT, Seymour Papert demonstrated how the discovery of scientific identity is rich in thought and feeling. His interest in my student object papers sustained me. This book is dedicated to him in gratitude.

For more than ten years, one of the courses in which I asked the question about childhood objects was Systems and Self, co-taught with Professor Mitchel Resnick of the MIT Media Lab. I thank Mitchel and all the students in Systems and Self for hours of conversation about objects and science. Special thanks are owed to Marina Bers and Michelle Hlublinka, who worked with Mitchel and me on that course and on developing ways to think about the object papers.

In 2001, the kinds of reflection represented in this book found a home in the MIT Initiative on Technology and Self, funded by a generous gift from the Mitchell Kapor Foundation. As a child of the Initiative, this book owes a debt to support from the Kapor Foundation and the Kurzweil Foundation, and to grants from the National Science Foundation and the Intel Corporation. The Spencer Foundation funded an Initiative conference on Adolescence and Technology that enabled me to develop many of the ideas represented in the introductory and concluding essays. My conversations with Courtney Ross and experiences as a mentor at the Ross School broadened my understanding of things and thinking. I am grateful for her friendship and those experiences.

At the Initiative, thanks are due to research assistants Olivia Dasté, Anita Chan, and Kelly Gray, who helped me read through the many hundreds of essays from which those collected here are a representative selection. In the early 1990s, Britt Nesheim was my first research assistant on what I have always thought of as the "objects and science" project. As it has developed, I have fondly remembered her enthusiasm and dedication. My assistants, Trude Irons and Judith Spitzer, made me and everyone around me more proficient and productive.

Kelly Gray remained involved with every aspect of the project as it was developed for publication. My debt to her is, as always, substantial and heartfelt. Howard Gardner and Mitchel Resnick read drafts of my writings in this volume. Every writer will understand the debt one owes to colleagues who are willing to grapple with ideas as they struggle to reach their final form. Nancy Rosenblum offered sage advice about the structure of the volume at the moment that I needed it most.

At the MIT Press, I am fortunate to work with a wonderful team that takes what I do and makes it better. My thanks go to Robert Prior for wise advice on the book's final organization, Deborah Cantor-Adams for meticulous editing, Erin Hasley for art direction, and Alyssa Larose for keeping the project on track. Colleen Lanick's enthusiasm for what I write makes it easier to write.

Watching how objects were enmeshed with the development of my daughter Rebecca's curiosities has always led me back to my students and their passions. As I complete this book, Rebecca is reading me the final volume of the Harry Potter series; its first books were ones I read to her when she was small. From Rebecca's reading I have learned that the wand chooses the wizard and the wizard chooses the wand, a perfect image for the magic that happens when young scientists find their object and their object finds them. For this and so many other lessons, I am grateful for what my daughter teaches me.

Sherry Turkle
Provincetown, Massachusetts
Fall 2007

Falling for Science

INTRODUCTION: FALLING FOR SCIENCE

Sherry Turkle

This is a book about science, technology, and love.

An eight-year-old sits braiding the hair on the tail of her My Little Pony doll, completely absorbed in the job. The shining plasticized hair is long and resilient; she plays with it for hours.

She starts by taking the tail and dividing it into three strands, which she braids together. Beginning again, she undoes that braid and divides the tail into nine strands. Then she braids groups of three until she has three plaits, which she braids together into one. Undoing this braid, the girl now begins with twenty-seven strands, braiding them first into nine, then into three, then into one. The girl is playing with My Little Pony but she is thinking about recursion.

This eight-year-old is one of my MIT students who, in this collection, write stories of their childhoods. What they have to say testifies to the importance of objects in the development of a love for science—a truth that is simple, intuitive, and easily overlooked.

There are many paths into science. This collection explores one of them, a path in which imagination is sparked by an object. It is about young people discovering objects that can "make a mind": a puzzle, a toy pony, a broken radio, a set of gears, origami. Here, three generations of distinguished scientists, engineers, and designers, and twenty-five years of MIT students in the course of their university training write about an object they met in childhood or adolescence that became part of the fabric of their scientific selves.[1] And since, for each of us, the many aspects of self are deeply enmeshed, these narratives about objects and science also explore themes of family, friendship, home, love, and loss.

In an ongoing national conversation about science education in America, there is a new consensus that we have entered a time of crisis in our relationship to the international scientific and engineering community.[2] For generations we have led; now Americans wonder why our students are turning away from science and mathematics—at best content to be the world's brokers, broadcasters, and lawyers, and at worst simply dropping out—while foreign students press forward on a playing field newly leveled by the resources of the World Wide Web. Leaders in science and technology express dismay. On this theme, Bill Gates stated flatly: "In the international competition to have the biggest and best supply of knowledge workers, America is falling behind." He also said, "In math and science, our 4th graders are among the top students in the world. By 8th grade, they're in the middle of the pack. By 12th grade, U.S. students are scoring near the bottom of all industrialized nations."[3]

When the Science Committee of the House of Representatives asked the National Academies, the nation's leading scientific advisory group, for ten recommendations to strengthen America's scientific competitiveness, the Academies offered twice that number.[4] There were recommendations to support early-career scientists and those who plan to become science teachers. There were recommendations to create a new government agency to sponsor energy research and to use tax policy to encourage research and development in corporate settings.

As sensible as these recommendations may be, they deal largely with financial incentives and big institutions. This collection suggests a different tack.

From my very first days at MIT in 1976, I found passion for objects everywhere. I had students and colleagues who spoke about how they were drawn into science by the mesmerizing power of a crystal radio, by the physics of sand castles, by playing with marbles, by childhood explorations of air-conditioning units. They also spoke of new objects. I came to MIT in the early days of the computer culture. My students were beginning to talk about how they identified with their computers, how they experienced these machines as extensions of themselves. For some, computers were "objects-to-think-with" for thinking about larger questions, questions about determinism and free will, mind and mechanism.[5]

Trained as a humanist and social scientist, I began to ask, what is the role of objects in the creative life of the scientist? What makes certain objects good-to-think-with? What part do objects take in the development of a young scientific mind?

The Collection

In the early 1980s, an MIT colleague, the mathematician and computer scientist Seymour Papert, wrote about how the gears on a childhood toy car had brought him into science. Fascination with those first gears made way for fascination with other gears. With practice, Papert learned to play with gears in his mind: "I became adept at turning wheels in my head and at making chains of cause and effect. . . . I remember quite vividly my excitement at discovering that a system could be lawful and completely comprehensible without being rigidly deterministic."[6]

The gears on the toy car brought Papert to mathematics, but more than an intimation of mathematics had brought Papert to the gears. They might have symbolized a connection to his entomologist father, a romantic but distant figure, who spent much of his time doing fieldwork in the South African bush. Seymour Papert's facility with gears might have been the first thing his father took pride in, and once this connection was made, Papert's object choice was overdetermined. We cannot know. What is certain in Papert's narrative is that thinking with and about things is not a cold, intellectual enterprise but is charged with eros. Papert says: "I fell in love with the gears."[7]

For over twenty-five years of teaching at MIT, I have made my first class assignment a question in the spirit of Papert's essay on gears: "Was there an object you met during childhood or adolescence that had an influence on your path into science?" Over the years, assigning my students a paper on childhood objects has sometimes provoked surprise from them, even anxiety. Students ask: "Why write about an object? Will I be able to find one?" I reassure these students that if they have trouble fixing on an early object, together we will find something appropriate for them to write about. No one will do poorly on this assignment. But then, once students begin to work, there are calls to parents to check their memories. There are conversations with siblings. My students go home for vacation and return to MIT with an object in tow. I typically devote one or two class sessions to reports on the objects of childhood; students have trouble keeping to their allotted times so we schedule extra meetings. Over the years, it has become clear that this assignment stirs something deep.

Here, I have chosen fifty-one essays from my collection of over 250 student essays gathered from 1979 to 2007 and followed them with essays on childhood objects by eight senior scientists, engineers, and designers—mentors who range in age from their forties to their seventies. Although the essayists' fields of interest cut across science, engineering, and design, the collection's title and my remarks refer to them to-gether as *science*. My focus is on what these fields have in common: a passion for the technical, for formal anal-ysis, for discovery, and for understanding how things work.

In the mentor essays one sees the arc of a life that takes a child from engagement with an object to scien-tific maturity. A boy is fascinated with the 173 steps of a hill in his hometown, by the stone terraces in his backyard, and by the wax hexagons of his beehives. He becomes an architect whose buildings celebrate the beauty of geometry. A hip high-school freshman in At-lanta has no interest in science until she discovers that it includes lasers, skydiving, and purple haze chemistry, their combined glamour drawing her toward a career in engineering. A child curious about the inside of a radio wonders what connects the circuits he can see and the broadcasts he can hear—concerns that will lead him to computational networks and questions about what links body and mind, brain and thought.

The mentor essays place us on a higher ground from which we can imagine where students may be heading, a prescience about pre-science. The mentor scientists were not given the student essays to read. Yet there are many connections across generations. Times have changed, certain objects have changed, but cu-riosity and a grammar of things and thinking have re-mained constant.

To illustrate something of that grammar, I ask the question, "What makes a scientist?" Activities with ob-jects provide some answers: building and sorting, play and vision, the way we use objects to model the world. Finally, there are the ways we take in the digital and the natural, what we program and what we sense. In the section on "building," I use one object, LEGO bricks, in a special way. Over the years, so many students have cho-sen them as the key object on their path to science that I am able to take them as a constant to demonstrate the wide range of thinking and learning styles that constitute a scientific mindset.

Thinking about scientists and their objects raises the question of how to best exploit the power of things to improve science education. Neither physical nor digital objects can be taken out of the equation; nor should either be fetishized. Over the past decades, we have seen an ongoing temptation to turn to computers to try to solve our educational crisis. It is natural, in a time of crisis, to avidly pursue the next new thing, but we need not lose sight of the things that have already worked. Awash as we are in new teaching materials (from smartboards to simulated science laboratories) object-play is not something to which today's teachers are necessarily attuned, although as early as third grade, young people interested in science can identify the objects that preoccupy them. Theirs are the minds we want to cultivate, but these students are often isolated, strangely alone with their thoughts.

One reason we don't pay enough attention to things and thinking is that we are distracted by our digital dreams; another is that traditionally, scientists have been reticent to talk about their object passions or, one might say, about passions of any kind. There was a canonical story about the objectivity and dispassion of scientific work and scientists stuck to it. In 1856, the essayist Walter Bagehot described the young scientist as an aficionado of the object world, yet Bagehot was ready to declare that scientists' involvement with "minerals, vegetables, and animals" spoke to an absence within their constitutions of an "intense and vivid nature." Scientists, he wrote, "are by nature dull and frigid and calm. An aloofness, an abstractedness cleave to their greatness."[8] In their autobiographical writings, scientists reinforced the idea that theirs was a discipline that faced nature with cool composure; lives in science were recounted in ways that separated reason and passion and saw objects through abstractions.[9]

But there has always been another story in which scientists' attachments to objects are red-hot. In recent years, this story is starting to be told.

Nobel laureate Richard Feynman begins his autobiography, *Surely You're Joking, Mr. Feynman!*, with a loving description of the "lamp bank" that he built when he was ten, a collection of sockets, bell wire, and serial and parallel switches, screwed down to a wooden base. Feynman plays with the lamp bank to get different voltages by setting switches up in different combinations, serial or parallel. He joyfully recounts his electronic universe: the radios he bought at rummage sales, his homemade burglar alarms and fuses. The fuses, made from tin foil, offer spectacle as well as intellectual excitement. Feynman sets them up with light bulbs across them so that he can see when a fuse has been blown. And he puts brown candy wrappers in front of the light bulbs so that a blown fuse translates into a beautiful red spot on his switchboard. "[T]hey would glooooooooow, very pretty—it was great!"[10]

It was the Great Depression, and Feynman's neighbors started to call upon the ten-year-old to fix their broken radios. In one case, Feynman figures out what is wrong with a radio that starts up noisily and then quiets down by asking himself the question, "How can that happen?" He lets his imagination move around the elements of the radio—thinking through the tubes, amplifiers, heat, RF circuit, grid voltages—and he comes up with a solution. The tubes are heating up in the wrong order. His neighbor's formulation: This child "fixes radios by thinking!"[11] In terms that Seymour Papert uses in his writing on education in the computer culture, the radios provided a "microworld" for learning.[12]

Feynman fell in love with electronics and, in the process, with thinking like a scientist. Like Donald Norman, who writes so movingly about radios in this volume, Feynman developed more than curiosity; he found a language for expressing it. He learned a certain way of asking questions and testing theories. For Feynman, radios carried ideas and made him famous in his neighborhood, an association of ideas and identity that is common to all of the essays in this collection.

Generations of Connection

Seymour Papert met his gears in 1930. The architect Moshe Safdie writes about a boyhood in Jerusalem later in that same decade; the cognitive scientist Donald Norman describes growing up with radios during the 1940s. Geologist Selby Cull met her chocolate meringue in the 1980s; computer scientist Andrew Sempere discovered the Holga Camera in the mid-1990s. This collection documents objects on the path to science, technology, and design over a seventy-five-year time span.

Over that time, there have been dramatic changes in the kinds of objects children have had presented to them. Yet in reviewing twenty-five years of science students' writing on their favored childhood objects, certain trends are apparent. One is an interest in transparency. Through the mid-1980s, MIT students who grew up in the 1960s wrote about radios, vacuum cleaners, wooden blocks, and broken air conditioners. These are things to take apart and put back together again. Students describe childhoods in which they fix what is broken or at least try to. They write about the frustration of not getting things to work but learning from their furious efforts.

By the end of the 1980s, my students begin to write about growing up with electronic games, lasers, video games, and "home computers," objects that are investigated through the manipulation of program and code. Yet even with the passage from mechanical to electronic, and from analog to digital, students express a desire to get close to the inner workings of their machines. The early personal computers made it relatively easy to do so. Machines such as the TRS-80, the Atari

2600, and the Apple II came bundled with programming languages and beyond this, gave users access to assembly languages that spoke directly to their hardware. Students write fondly about programming in assembler and of the pleasures of debugging complex programs. Metaphorically speaking, an early personal computer was like an old car in your garage. You could still "open up the hood and look inside."

However, by the 1990s the industry trend was clear: digital technology was to become increasingly opaque, reshaped as consumer products for a mass market. The new opacity was cast as transparency, redefined as the ability to make something work without knowing how it works. By the 1990s personal computer users were not given access to underlying machine process; computers no longer arrived with programming languages as a standard feature. Beyond this, programming itself was no longer taught in most schools. Even so, young people with a scientific bent continued to approach technology looking for at least a metaphorical understanding of the mechanism behind the magic.

Beyond seeking a way to make any object transparent, young people across generations extol the pleasure of materials, of texture, of what one might call the resistance of the "real." In the early 1990s, computer scientist Timothy Bickmore's experiments with lasers, "passing the laser through every substance that I could think of (Vaseline on slowly rotating glass was one of the best)," recall the physical exuberance of Richard Feynman's candy-wrapped light bulbs of a half-century before. For Selby Cull in 2006, geology becomes real through her childhood experience of baking a chocolate meringue: "Basic ingredients, heated, separated, and cooled equals planet. To add an atmospheric glaze, add gases from volcanoes and volatile liquids from comets and wait until they react. Then shock them all with bolts of lightning and stand back. Voilà. Organic compounds. How to bake a planet."

Cull's joyful comments—"Voilà. Organic compounds. How to bake a planet"—introduce the most well-known of the intergenerational experiences described in these essays: the moment of scientific exultation, the famed "Eureka" moment of raw delight, here consistently recounted with an object as its focus. In the 1950s Donald Ingber learns to anticipate the thrill of the gestalt with his color-by-numbers pencil set when "after coloring in multiple scattered spaces, I was always elated when I penciled in that key space that caused all the other colored tiles to merge into a single coherent image." In the 1980s Jennifer Beaudin's realization that she will change but her house will stay the same compels her to map its stable contours. In doing so, she comes to another discovery, one she finds even more startling: "A wall of one room could be the wall of another. . . . Indeed, all the rooms were adjacent to each other and formed a whole. I can remember how it felt to suddenly see something new." In the early 1990s while fishing with his father, Cameron Marlow looks up at the motion of his fly line and is reminded of drawings of long, continuous, flowing lines he had made in algebra class. "I realized that the motion of my hand had a direct effect on the movement of the line, much in the same way that the input to a function produced a given output. Without any formal understanding of the physics involved, I was able to see the fly rod as representing a function for which I was the input. . . . From this point on, the fly rod was my metaphor for understanding function in mathematics."

The themes that cut across generations introduce those of the collection as a whole. Objects provide encounters with transparent systems and manipulable microworlds. They provide opportunities to develop intimacy with objects and to develop a personal thinking style. Finally, objects provide occasions for young people to make the most of the analog and the digital, the natural and the simulated. Objects are not the only path into scientific creativity, but they are one powerful path. In these pages, we see objects helping young people realize something about self and something about science, keeping in mind that to realize means to understand and to build.

I have asked my students to consider objects at a time of transition, or as my colleague Nicholas Negroponte would put it, at a time when the world was moving from atoms to bits.[13] I have traced students' object passions across the years of the digital revolution and found that in the end there has not been so much a migration to a new digital world but rather that children now grow up in many worlds. They are seduced by the control of the digital, the freedom of the virtual, but always brought back to the physical, the analog, and of course, to nature.

Transparency

Neuroscientist Susan Hockfield begins her essay on microscopes with the question, "How do you understand how something works? From as early as I can remember, I wanted to see inside things, to understand how they worked." Cognitive scientist Donald Norman echoes the sentiment when he describes taking apart a radio:

> I loved the insides of the radio. I can remember the undersides, a mesh of thick wires running this way and that, covered with dust and cobwebs, connected at junctions with nice dull solder balls, with multiple large cylinders connected to them. . . . The radio transformed my life. I finally had focus: to understand the hidden mechanisms of electronics.

Even after Norman moves from electrical engineering to psychology, he feels that he is pursuing the same goal, understanding "mysteries of hidden, invisible mechanisms, but now my focus was on the human mind rather than the electronic circuit."

A half century later, MIT student Kwatsi Alibaruho has the same exuberance for the insides of telephones. His mother provides him with "three old telephones, a small wrench, and a screwdriver. . . . I took to taking apart the telephone—something I loved!" Prior to this, Aliburaho had spent many months building toy telephones out of interlocking LEGO bricks. He feels that the LEGO telephones prepared him for getting to work on the real thing. Building begets a love of building. For this builder, what thrills most is what is most hidden. Alibaruho comes alive when the invisible is made visible. Of the telephone he says:

> Seeing all of the wires and screws inside was an incredible high. . . . I spent hours engaged with my phones. My goal was simple. I wanted to take the phone apart and then put it back together. . . . Finally, I could take a telephone completely apart and put it back together so that it actually worked. I did not know how the components worked, but I began to get a feel for how they fit together.

From LEGOs to telephones, from telephones to bicycles, Alibaruho likes to "disassemble and reassemble." For him, the fun of a new bicycle was not riding it, but taking it apart and putting it back together, over and over again. After a time, assembly and reassembly becomes its own pleasure, a kind of meditative activity. Living in the worlds of his constructions, Alibaruho works with his objects in the spirit of what French anthropologist Claude Lévi-Strauss calls *bricolage* or tinkering, the combining and recombining of materials on their way to becoming scientific thought.[14]

Norman and Alibaruho were involved with objects they held in their hands, things exemplary of what Lévi-Strauss, punning in French, called "goods" to think with, *bons à penser,* that are also "good to think with." Hockfield's narrative adds another dimension. She plays with objects of her imagination that are none the less concrete for not being in her physical grasp. Hockfield can't remember ever being without a magnifying glass to help her get inside of things: a door latch, a watch, an iron, a toaster, a fan. But she also played with the objects of her dreams. She loved elaborate miniatures, trains, and dollhouses, but didn't own the kind she dreamed about, the most complicated kind, "with electricity, running water, a heating system, plumbing, all of the mechanics. I wanted a fully functioning miniature, so that I could understand how a house works." Papert, Alibaruho, and Feynman handled mechanisms, made them transparent, and ended up working them in their minds. Hockfield shows us the power of objects imagined as transparent from the very start.

Microworlds

Alibaruho describes mechanical constructions that draw him into worlds he can create and control. As he gains fluency with his objects, they become elements for building physical systems and for building his mind. "I thought of my imagination as constructible."

The metaphor of building is central to the Swiss psychologist Jean Piaget's constructivist description of child development.[15] Its basic tenet: children build theories based on the objects they meet in the world. Seymour Papert was one of Piaget's students and saw the relationship of object to theory in more activist terms, closer to the experience Alibaruho describes. Papert moved to a constructionist position: children make their minds through actual building. In Papert's model, we can expect that if we give children new materials, they will build different things and be able to think new thoughts. Piaget's constructivism takes the object world as something of a given; constructionism puts the child on the prowl for new objects, new ideas.[16]

In Moshe Safdie's essay, we see a child looking for new materials, seeking the joy of construction. Drawing limits him. He needs to build:

> Though I could draw with considerable ease, drawings seemed inadequate to describe what I had in mind. I joined with a friend and we decided to make a model. With a model, we could create a lake formed by a dam, show the water drop into turbines below, set windmills on the ridge, irrigate the terraces downhill, and so on. In the basement of a building that had been used as an air raid shelter during the Second World War, we found an old, unused door and used it as the base for our model. We purchased many pounds of clay to form hills, a lake, and valleys. We cut up little weeds to represent trees. We used dyes to color the landscape. We tried simulations by pouring water above and seeing it trickle downwards and we began searching for a pump that might keep the system going.

Richard Feynman's radio connected him to proud parents and became an offering to a wider community. Safdie's model provided a similar opportunity for social success. In both cases, construction nurtured social identity. Designer Sarah Kuhn's models have a different emotional valence: they are her private retreat.

Kuhn takes a set of wooden blocks and builds a fortress. She analogizes her blocks to the virtuous objects created by the nineteenth-century German educator, Friedrich Froebel, the inventor of kindergarten. Froebel proposed a set of twenty objects, "Froebel's gifts," each designed to impart specific competencies.

Like the gifts, Kuhn says that her blocks are powerful teachers; her blocks-world is the perfect place to learn how to think like a designer—how things fit, how structures work. But beyond this, Kuhn's blocks-world provides a safe haven, "my private universe." After her brother is born, her parents divide up the attic and she gets her own room. It is in this room where she sets up her blocks. Her bed is at the center of her world; it is her safety raft. She extends it with adjacent blocks.

> The scope of my ambition now expands to fill the room; the bed and the floor become part of the action. My grandfather builds me a table in the form of a giraffe, and I incorporate it, too, into my constructions. Usually my bed is a raft. . . . Anything touching my bed is part of the raft, keeping me and my stuffed animals safe and dry.

Kuhn, both client and designer, uses her constructed world to address the anxieties of childhood. She assuages fear by playing with her worst ones, declaring the blue carpeting of her attic bedroom to be a hostile sea, "its unplumbed depths harboring countless marauding sharks, the bogeymen of my childhood." The family lives only a few miles from Alcatraz, the island maximum security prison, and Kuhn "shiver[s] at stories of would-be escapees who come to a bad end. As I extend my construction, I extend my world of safety."

Kuhn's narrative exemplifies the integration of thought and feeling in a learning microworld. She is learning about design and she is saving herself, all at the same time. Children bring their emotional needs to their intellectual constructions. Outside, the world is complex, parents are occupied with grown-up matters, a new sibling presents competition. In the blocks world, Kuhn is self-sufficient. Kuhn shows us how blocks can offer a "just-right" emotional fit. Her blocks-world problems are ones she can solve. The blocks-world enables her to construct just the right degree of separation from her siblings, from her parents, from too-big problems. The "just-right" fit of Kuhn's physical microworld helps us to better understand today's children and their virtual microworlds. They play video games that are "just-right" in their presentation of increasingly difficult sequences to master, sequences just as comforting as blocks challenges are to Kuhn.[17]

Alibaruho notes that he becomes lost in his design microworlds. When he immersed himself in a bicycle-making microworld, he thought about everything in terms of bicycles and "The bicycles that I worked on felt like parts of me." Kuhn, too, becomes what she builds. She is not using her blocks to build a model but a world to her scale: "Back in the playroom, I would have had to imagine myself an inch high to live in my blocks compound; in my bedroom, I can be my own size. I inhabit my construction world." The objects of her microworld have taken on a special physical intimacy.

Object Intimacy

Stereotypes about scientific work would have scientists, engineers, and designers thinking through problems in a "planner's" style, a top-down, divide-and-conquer style in which objects are kept at a distance. Of course, some scientists do use this style, and some use it most of the time. Others describe a hybrid style that moves back and forth from top-down planning to a more fluid and experimental bricolage, or "tinkerer's" style, one that is likely to leave more space for object intimacy. Yet the planner's style became frozen in the public imagination (and to some degree, the science education community's as well) as the way one does things in science, and even more broadly, what it means to think like a scientist.

Historian Evelyn Fox Keller writes about scientists' resistance to acknowledging the intimacy of their connections to objects. She sees its roots in a male-dominated view of mastery that equates objectivity with distance from the object of study.[18] In contrast she describes the attitude of a Nobel laureate, the geneticist Barbara McClintock. For McClintock, the practice of science was a conversation with her materials. "Over and over again," says Keller, McClintock "tells us one must have the time to look, the patience to 'hear what the material has to say to you,' the openness to 'let it come to you.' Above all, one must have a 'feeling for the organism.'"[19]

McClintock talks about the objects of her science, neurospora chromosomes, in terms of proximity rather than distance, in terms that recall what we have heard from Kuhn describing her blocks and Alibaruho talking about his telephones. The chromosomes were so small that others had been unable to identify them. But the more McClintock worked with them, "the bigger [they] got, and when I was really working with them I wasn't outside, I was down there. I was part of the system. I actually felt as if I were right down there and these were my friends. . . . As you look at these things, they become part of you and you forget yourself."[20] Similarly, when Susan Hockfield describes the pleasures of the electron microscope, she uses a language that puts the emphasis on "being there."

> The microscope itself was a large vacuum chamber, with an array of pumps to evacuate it. The electron beam was projected through the sample under study onto a screen, which had to be viewed in the dark. So the experience of using an electron microscope, in a darkened room with lots of noise from the vacuum pumps, felt very much as though you, yourself, were "in the microscope." I would spend hours in the microscope, scanning tissue, with a wonderful feeling of being inside the specimen.

Hockfield's intimacy with the microscope, McClintock's being "among" her cells—both evoke what the psychoanalyst D. W. Winnicott called the "transitional object," those objects that the child experiences both as part of his or her body and as part of the external world.[21] As the child learns to separate self from its surroundings, the original transitional objects are abandoned; one gives up the prized blanket, the teddy bear, the bit of silk from the pillow in the nursery. What remains is a special way of experiencing objects that recalls this early experience of deep connection. Later in life, moments of creativity during which one feels at one with the universe will draw their power from the experience of the transitional object.

For Keller, any description of scientific practice in which the scientist is distanced from his or her object of study cannot stand alone. It needs to be seen in relation to other descriptions such as those provided by a McClintock or a Hockfield that are about intimacy and presence. Keller takes the exploration of this second style as part of a feminist project in science but makes it clear that there are many male scientists who work in a "close-to-the-object" style. The scientific culture, however, has made it hard for them to talk about it or, perhaps, even to recognize it for what it is. But once young male scientists are asked about their objects, they offer rich evidence of such intimacies. One of my students spoke to me about translating the tactile experience of playing with marbles to feeling the laws of "physics in his fingertips"; another tells me that as a child he was so involved with a carpenter's ruler that it became a physical template for intuitions about proportion. Even as an adult mathematician, when he divides a number by two or four, he sees the ruler collapsing in his mind. In this collection, Thomas P. Hermitt writes about diving deep within a prism for inspiration, shrinking himself, as did McClintock, to its scale in order to make his body feel at one with its structure:

Visualizing waves of light bouncing off nuclei, slithering through electron clouds, and singing across the vacuum between the stars became an obsession. I never tire of leaving the ordinary, everyday world, shrinking myself down to the size of an electron and diving headfirst into my prism where a front row seat for the spectacle of nature awaits.

In common with these examples of what Papert would call "body syntonic" relationships with objects, Austina De Bonte finds herself thinking with her fingers when she learns to build *siaudinukai,* an old Lithuanian folk craft. *Siaudinukai* are three dimensional objects made by threading straws on string. No Froebel gift was ever more evocative than De Bonte's thread-and-straw world. As an adolescent, De Bonte goes to a Lithuanian summer camp to explore her ethnic identity. There, she uses *siaudinukai* for an allied exploration: the search for rules for a "stable" structure, one that is "rigid, reasonably strong, and structurally complete." At camp, she gives all these things a name: *solidness.*

I discovered, mostly by example and through trial and error, that I couldn't make a solid structure that wasn't based on triangles. I also found that every link in a *siaudinukas* was vitally important—the structure was often fully collapsible and foldable right up until the very last straw was secured. Furthermore, I discovered that this was actually the mark of a good structure—if the *siaudinukas* was rigid before I was done executing my plan, then quite likely I had redundancies in my planned structure that were not only unnecessary but in some cases actually caused the structure to lose its pleasant symmetry along an axis, hang crooked, or put unwanted tension on other straws.

The year she wrote her essay on straws, I watched De Bonte run a small workshop on how to build *siaudinukai*. She brought straws and thread and Lithuanian snacks. One rapt five-year-old was always pleased when she could get an early-stage structure to stand on its own; this made it easier for her to thread the straws. At the workshop, De Bonte was gentle and firm in her rebuke: "If it looks ready too soon, it's not ever going to stand on its own." The lesson needed to be repeated three times. Each time De Bonte made her point, it seemed to have a wider meaning. In the end I was moved by what seemed its most general meaning: suppleness is the precursor to what is ultimately most secure.

Personal Thinking Styles

The strategies children use to engage with objects can be categorized in two ways. A first focuses on stages of development, and a second looks to personal styles. In the first, the developmental framework, we can imagine three stages: metaphysics, mastery, and identity, each of which provides its own bridge to scientific curiosity. In the metaphysical stage, objects help children consider basic questions about aliveness, space, number, causality, and category, questions to which childhood must give a response. Children wonder at objects; objects provide early inspiration for the child scientist. Here, we think of Papert and his gears, comtemplating the basic rules of causality and sequence, and a young Britt Nesheim, exploring the mysteries of number and size with her toy mailbox.

In the mastery stage, which begins at around age eight, children use objects to prove themselves and their ability to control the world. At this stage, children's thoughts often turn to winning. By the time he is nine, maps and routing offer Steven Schwartz "a world of mastery and control." Of course, sports and social life can also provide material for developing feelings of control and mastery, but objects tend to offer experiences that produce the greatest certainty. You can lose a baseball game. But with practice, you will always be able to sort objects, plan a route, or put a disassembled telephone back together.

The seduction of what can be precisely known and controlled can be a path into science. Object mastery can also provide opportunities to make false starts without penalty, to recast "getting it wrong," as a step on the path to "getting it right." During the mastery phase, this object optimism, explicitly called "debugging" when students talk about programming, is part of the positive experience of other objects as well.

Finally, with adolescence, children's concerns turn to identity, and objects help them become who they are.[22] Chuck Esserman writes that when he was developing his identity as a techie, he thought of himself as his bike. Rosalind Picard wants to be the kind of girl who has a yellow nonregulation notebook and who experiments with lasers and exotic chemistry.

Metaphysics, mastery, and identity are developmental stages, but no handle cranks, no gear turns to graduate a child from one stage to the next. No stage is ever fully completed; we continue to work on all of these issues throughout life. And we do so with our objects in mind. This is so much the case that metaphysics, mastery, and identity are as much styles of engagement with objects as they are stages of development. We saw how in Kuhn's and Alibaruho's early engagements with objects, they were most concerned with control; later, identity—seeing themselves as builders and designers—takes center stage.

This same mix of mastery and identity, stage and style, operated for De Bonte. The *siaudinukai* show her what she can do but also who she is—not only as a Lithuanian but as a thinker. Working with *siaudinukai* concretized her personal thinking style.

De Bonte's style is to play with her materials until she gets things right. She is a classic bricoleur. She tries one thing, steps back, and tries another. As she puts it:

> Sometimes I would just start stringing some straws together, looking for ideas; once something took shape, it was easy to find ways to extend or elaborate on it. . . . Often I wouldn't be able to tell for sure whether a complicated structure would be solid until putting in the very last piece.

Those who predesign their straw structures do not do better work than she. They simply have a different style.[23] From a pedagogical point of view, looking closely at objects leads us to greater respect for the many ways of thinking like a scientist, engineer, and designer. In this collection, this range of styles is dramatized by narratives such as that of De Bonte and, as a group, by the range of responses to LEGOs, named by many of my students as a crucial object on their path to science.

Some children use LEGOs to create highly realistic structures. For others, only fantasy buildings hold any interest. Some maximize their LEGO resources by constructing hollow buildings to conserve bricks. Others challenge themselves to use all their bricks in one structure, no matter how baroque the result. Some follow a plan of detailed instructions; others throw instructions away. Some keep their constructions as trophies; others destroy what they have done as soon as it is completed. Some build designs they can live in; others build for fantasy characters. For one child, the most exciting thing about LEGOs is the LEGO "bump"—its unity suggests the idea of an indivisible particle; for another, building with the bricks is less exciting than classifying them. He expresses his creativity when he contemplates the range of algorithms possible to sort the colors, shapes, and types of LEGO bricks. Children's experiences with LEGOs dramatize that the choice of an object does not determine its creative impact. These children, young scientists all, make objects their own in their own way.

For Alan Liu, who builds structures for richly imagined LEGO people, "Most satisfying to me was that each member of the space colony had a personal identity. I had men and women who had marriages and children." In Liu's medieval LEGO world, the king "was a fool . . . the queen disliked her silly husband so much that she spent more time with the prince." Liu makes his LEGO buildings actors in his characters' dramatic lives. When the foolish king battles the astute leader of the space colony, the king always loses because he does not have the wits to work around a design flaw that Liu has built into the monarch's LEGO equipment. But while "the king never worked around his weakness[,] the spacemen always exploited it." When Liu is about nine, his relationship to LEGOs changes. Instead of building for his characters, Liu begins to build for himself. He takes apart everything he has built and starts from scratch: "I don't know if I was playing any more. It felt like serious designing."

While Liu is happy playing with LEGOs that come in space and medieval kits, Sandie Eltringham doesn't like kits at all. She is interested in LEGO as a protean resource. Eltringham focuses on the sample pictures on the LEGO box lids because they come *without* instructions. She looks for building clues by analyzing the shadows that the suggested structures cast on the walls in their photographs. Eltringham takes LEGO people out of her play. Her only use for them is as gauges to help her build furniture and cars to scale. For Eltringham, nothing can compete with hearing "the loud snap of two pieces correctly put together" as she creates a perfect miniature. As much as putting things together, Eltringham enjoys taking things apart, sorting each piece by color, shape, and type. In the end, her passion is the pleasure of classification.

The Analog and the Digital, the Natural and the Simulated

Electrical engineering student Mara E. Vatz begins her narrative about vacuum tubes with a declaration of love for a fifty-year-old Magnavox record player being tossed out on the street. She feels compelled to rescue it: "The Magnavox took hold of me." By the time of her schooling, electrical equipment did not reveal its circuitry to the naked eye and hand. Circuits were stamped on chips, no longer traceable through their wiring. Frustrated by the opacity of contemporary electronics, Vatz finds what she is looking for when she meets the transparent Magnavox. She says, "I turned the whole thing upside down to get a peek at the circuitry inside."

Once "inside" the Magnavox, Vatz discovers its treasure. It is built with analog devices: "[M]y Magnavox had . . . vacuum tubes instead of transistor amplifiers." Vatz is not so much thrilled by what the vacuum tubes can *do* (carry sound with less potential noise) as by what they *are*. In an opaque, digital world, they embody transparent, analog knowledge. Vatz is resigned to the fact that the fragility, size, and expense of vacuum tubes dictate their disappearance from modern electronic equipment but is upset that the ideas they carry are being lost as well. She finds it nearly impossible to get information about the vacuum tubes. Vatz is majoring in electrical engineering, but her professors talk about vacuum tubes "only when [they] reminisced about them, the way they might about old friends—brilliant and wonderful friends—that we, students born too late, would never have the chance to meet." Current books don't mention them at all. Vatz begins to seek out old books, very old books, and finally finds one of her grandfather's engineering books where vacuum tubes are discussed and illustrated. Vatz is fascinated to find that in the old textbook, the vacuum tubes are put in the context of the history of the science from which they arose. For Vatz, meeting vacuum tubes leads to an interest in the ideas they carry and to the notion that if objects are lost, ideas can be lost as well. Reclaiming an object means reclaiming a set of ideas.

The generations of students who have grown up in the post-analog world can understand Vatz's concerns. They experience the trade-off between the kind of transparent understanding offered by the world of mechanism and the heightened sense of control offered by the increasingly complex and opaque digital realm. In digital culture, you may not have a traditional understanding of how things work, but you have enormous power to make things happen, to invent new possibilities, to find new creative outlets. Control and understanding begin as competing parameters, but at a certain point, control begins to feel like a kind of understanding. It is the way of understanding in digital culture. With digital objects, with programming, one is free to build "straight from the mind."

"Building straight from the mind" are the words of an MIT student I call Anthony, a self-identified computer hacker of the 1980s, who had grown up in the analog world.[24] As a young boy, Anthony took clocks apart and "tried to put them together in new ways—to make new kinds of clocks."[25] But there were limits to how much Anthony could make clocks into something new. When he met computers and programming, he sensed that he was in a world with no such limits. In a programmed microworld, the laws of gravity need not apply. "When you are programming," says Anthony, "you just build straight from your mind."

> Why do you think people call their ideas brainchildren? They are something you create that is entirely your own. I definitely feel parental toward the programs I write. I defend them and want them to do good for the rest of the world. They are like little pieces of my mind. A chip off the old block.[26]

Anthony's pride in his brainchildren is expressed in the object aesthetic of digital culture.[27] It is an aesthetic that parries constraint and claims to be a universal language while enabling individual expression.[28]

Here I have used the word *microworld* as a term of art. In the 1960s Seymour Papert used it to describe software that was designed to bring ideas from computation into the thinking vocabularies of children. Papert, influenced by Piaget's ideas about the child as scientist, wanted to broaden the scope of what children could be scientific about. Piaget had argued that children use the objects of everyday life to develop theories about such things as space, time, number, and causality, as well as what it means to be alive and conscious.[29] Papert underscored that children develop intellectual fluency about such questions because the world provides concrete materials with which they can think them through. Papert reasoned that if thinking about aliveness could be developed in a world of living and not-living things (call it Lifeland) and fluency in French could be developed in Frenchland (say, for example, in France), then thinking about mathematics might flourish in Mathland. He believed that this was best thought of as a world that could be built within the computer, a mathematically based machine in which programming languages would construct worlds that operated by mathematical rules.

But Mathland needed a way to connect to people. For Papert, that was the turtle, a robotic creature connected to the computer that took its marching orders from the Logo programming language. In that language, an original robotic "floor turtle" evolved into a "screen turtle," a triangular cursor on a computer screen. Both floor and screen turtle could be told where and how to move and whether or not to leave a "trace" by lowering a pen, physical or virtual. So a turtle could be made to trace a square by giving it the following sequence of Logo commands:

```
PEN DOWN

FORWARD 10

RT 90

FORWARD 10

RT 90

FORWARD 10

RT 90

FORWARD 10

RT 90

PEN UP
```

The turtle had many virtues as a way to convey ideas, foremost the fact that it connected with both mind and body. Children wrote the programs that controlled the turtle—that was the mind connection—and children could "play turtle," directly identifying their body with that of the turtle as a way to learn programming. In a classic assignment, Papert would ask a child to program the computer to make a circle. The most direct way to solve this problem is to "play turtle" and act out the drawing of a circle. Using the "body-as-turtle" method, children could come up with the Logo program "To Circle." This program instructs the turtle to go forward one unit, turn by one unit, and then repeat this again and again. By trial and error, children learn that 360 repetitions is the right number to get you to a full circle. Whether the turtle was a floor robot or a point on a computer screen, the body-to-body path to mathematical insight remained intact.

This kind of thinking—ask digital objects to relate to body and mind, ask digital objects to carry ideas—continues to inspire research on educational computing in the tradition of Papert's Logo work. At MIT, Mitchel

Resnick, once one of Papert's students, leads a research group on science education under the rubric of "Lifelong Kindergarten." In a personal introduction to his work, Resnick explains his fondness for the kindergarten metaphor as follows:

> Kindergarten is one of the few parts of our educational system that really works well. In kindergarten, kids spend most of their time creating things they care about: building towers out of wooden blocks, making pictures with finger paint, creating castles in the sandbox. As they playfully create and experiment, kids begin to develop new understandings: What makes structures fall down? How do colors mix together?[30]

After kindergarten, says Resnick, education too often shifts to "broadcast mode," with schools trying to deliver information, rather than allowing students to learn through building and experimenting.

Resnick's research group has made "programmable bricks" for LEGO construction kits that enable children to program LEGO objects much as one could program Papert's floor turtle. Only now, the objects are not only able to move, they are able to sense their environment and communicate with other objects in their world. In this way, LEGO constructions evolve into LEGO creatures. Papert's Logo world enabled children to explore mathematics and physics. Resnick's designs enable children to explore biology and psychology. They can investigate notions such as feedback and emergence (how complex behaviors can emerge from simple rules) that were previously viewed as too hard for children to think about.

Indeed, one of Resnick's central ambitions has been to bring ideas about the importance of decentralized control into the thinking vocabulary of children.[31] In arguing for its importance and in pointing out the resistance to it, Resnick cites the work of Evelyn Fox Keller.[32] When Keller's research led her to propose a decentralized model of cellular communication, she found the community of professional biologists arrayed against her in almost-universal protest. They preferred thinking of communication within cells as a centralized, top-down, command-and-control structure.[33] Resnick creates programming environments that dramatize the power of decentralized phenomena. In one of these, the programming language StarLogo, objects can be made to respond to changes in their local environment. Over time, if each object follows simple local rules, global patterns emerge that appear to have been designed from "above." The effects are stunning, memorable demonstrations of what can be achieved by decentralized processes. They help to fight our prejudice that all structure is planned. Structure can also be emergent. The demonstration is timely since it is at the heart of embattled evolutionary theory.

Most recently, the Lifelong Kindergarten group has focused on integrating computation into children's everyday experience. It has developed a new language, Scratch, in which one is able to use popular culture— music, art, storytelling, and video—as elements in a program. Resnick's work illustrates the best in digital media for education. It takes the Froebel gifts as its touchstone and uses them as inspiration to enhance each child's creativity. Unfortunately, many digital worlds make everything possible but constrain experience, a trade-off apparent in most science classes where virtual science laboratories are in use.

Experiments in simulated laboratories are usually made to work out. The experiments become a performance; students are elements in that performance, but the main actor is the program. One can program random failures into the simulation (spilled coffee, broken equipment, an overheated atmosphere, the delivery of defective mice), that is, the kinds of things that make laboratory science go awry. But these slips will not *happen,* they will be *found.* They are there from the start, placed in advance by a programmer in charge of the virtual world.[34]

The young scientists in this collection show a certain craving for the contingent, for that which cannot be anticipated. Today's complex computer games go to considerable lengths to "program in" surprise (the famous "Easter eggs" within computer games that offer special tips and bonuses), but even the determined children who find them are not making the kind of discoveries that give an experience of full ownership. Someone has been there before: the programmer.

Consider a computer project that uses computational straw to design and build virtual *siaudinukai.* There, the lesson of "seeming solidness" might emerge from thousands of iterations of a building program that models virtual straw. A user manipulates plastic dowels that are represented in the computer. When a builder makes a simple structure, the computer transforms it into thousands of alternate configurations. Thus programmed, building *siaudinukai* could take place on a vastly wider scale. Robust structures would pop up from the many virtual configurations developed in collaboration with the computer. For some students, such multiple iterations of geometrical possibilities would facilitate learning.

Contrast this power with how De Bonte learned the lessons of the *siaudinukai*—through her fingers and in community with her peers, learning as part of her contact with Lithuanian culture. For some people, what might have been magical about the straw shapes will be lost in their digital variant. Otherwise put, in a digital world, children may get the point, but that may be all that they get.

Or contrast the exploration of the principles of heat and energy in a computer world with the adventure recounted in Daniel Kornhauser's essay on discovering a heat source within his shirts. The seven-year-old Kornhauser begins with a real problem that involves his comfort, his family, and economic realities: after a move from Mexico to France, Kornhauser is freezing cold at night, but it is beyond his parents' means to purchase twenty-four hour heat for the new family apartment. Kornhauser becomes fascinated by the sparks emitted by his thermal underwear. He decides he has "tapped into an undiscovered source of energy." It is his discovery, his "secret idea": "I shared most of my ideas with my father, but this one I kept to myself. . . . I was sure that I had developed a way to generate electricity that would enable us to keep the heat on all night long."

Ultimately, Kornhauser's father talks to his son about the scientific principles that deflate his first fantasies. But in the process of making his discovery, Kornhauser has become committed to science. That commitment is not tied to things going right. Indeed, Kornhauser says that although things went wrong, what matters is that he owned his failures. Kornhauser says that he uses his failed projects to think about entropy and the conservation of energy. As his mastery over large scientific principles grows, the world becomes reanimated, luminous. He learns that science leaves room for "invisible things," that "magic was not only to be found in fantastic tales; you could find it everywhere, invisibly surrounding you."

For the three-year-old Matthew Grenby, a bottle of soap bubbles also contains such magic. The soap bottle is a "made" thing, but Grenby experiences it as part of nature:

> I would shake the bottle and thousands of small bubbles would fill the small airspace in the container. Once agitated, any amount of continued shaking would have no effect, except to reshuffle the existing bubbles. And then, somehow, impossibly, the bottle would lighten. The fluid would

disappear among the bubbles. If I wanted to play again, I had to wait until the bubbles turned back into fluid. I have no recollection of ever attempting to unscrew the bottle's cap. In spite of my frustration, I was content with the integrity of this little world. The bubbles offered no explicit insight into the ways of their world; rather they left an almost imperceptible impression.

The soap bubbles are only one natural object that draws Grenby to science. There is also an avalanche and a creek, sometimes dammed up with mud and twigs. All of these inspire him to feel awe for the mystery of science and at one with all that is beyond him. He calls it "a humility born of the violence of the mountain."

Nature encourages us to be messy because it is. The geologist Selby Cull thinks of it as the "challenging, beautiful, and delightfully tasty." When we deal with nature, we need to become comfortable with the idea that things may go unresolved for a while, that we may break things that are not easily replaceable, and that actually, things may not work out. In simulated science, there doesn't have to be any waiting. Time can be sped up. And when something breaks, the simulation can be run again and whatever was broken can be magically restored. Simulations encourage the idea that one can push forward to resolution—of the game, the quest, the experiment. One can push forward because possible resolutions are already there, in the program.

While the digital is explicit, the physical can exhibit a certain reticence. In this collection, among the most dramatic stories of learning is one that has as its central actor a piece of furniture whose presence is never quite acknowledged. This is a table in Alan Kay's fourth-grade classroom, an old dining table belonging to his teacher, Miss Quirk. It is completely covered, as Kay puts it, "with various kinds of junk: not only books, but tools, wires, gears, and batteries." Miss Quirk never mentions the table. But students are drawn to it. On it, Kay finds a book about electricity and the materials he needs to follow through on one of the projects in the book:

> One afternoon during an English class I set up my English book with the smaller electricity book behind it, and the large dry cell battery, nail, wire, and paper clips behind that. I wound the bell wire around the nail as it showed in the book, connected the ends of the wire to the battery, and found that the nail would now attract and hold the paper clips!
>
> I let out a shriek: "It works!"

Much more is working in this story than Kay's new electric magnet.

The objects on Miss Quirk's table are presented as bits and pieces of things. Students are given space and time to discover which objects should belong to them. Students can take things off the table without knowing why they are doing so. Kay gets a result, but he didn't go to the table with the aim of achieving one. The objects on Miss Quirk's table don't have predetermined uses. A beaker can be used to pour chemicals, water plants, hold flowers, or make a vacuum. The table presents familiar objects to get you to unfamiliar places. Kay is first drawn to a book because he thinks of himself as a reader, but he ends up making a circuit.

Miss Quirk's table is one thing; Miss Quirk is another. Her presence means that she can see the young Alan Kay slouched down, trying to hide his electrical circuit. She intervenes to encourage him; she intervenes to build a relational bridge between an object passion and the rest of life. In this collection we meet parents, relatives, friends and teachers who bring children old telephones, maps, LEGOs, and blocks. Sometimes they work alongside children. Sometimes they appreciate what children have accomplished. Like Miss Quirk, they bring sociability and community to what begins as a private moment. It is perhaps less important to think about the dichotomy between physical and digital artifacts than to make sure that we communicate with our children about their objects—physical or virtual. Virtual objects challenge us to learn how to enter someone's digital world as easily as moving aside the book that the young Alan Kay used to hide his circuit.

Yet Kay himself sees Quirk's table as close to a programmed object. For Quirk chose her materials carefully so that the children in the class would come upon objects in the right sequence. Kay says, "Because discovery is difficult, children have to be given scaffolding for their ideas. They need close encounters with rich materials; they need a careful yet invisible sequencing of objects." For Kay, the table is an inspiration for educational programming. He aspires to build computer microworlds in which, like the worlds designed by master teachers, students are not told what to learn but are encouraged to explore in sequences that "will enable them to make the final leaps themselves."

Kay's aspirations for the computer's use in education—as a laboratory for exploration—is close to how scientists use the computer. Chemists manipulate virtual molecules; biologists fold virtual proteins; physicists explode simulated nuclear devices. The children of today, the scientists of tomorrow, need to be comfortable in virtual space. Children's passions for objects teach us that there are possibilities in the digital that should be pressed into the service of a more effective science education and there are things to learn from the physical that are worth fighting for.

The Things That Work

Science is fueled by passion, a passion that often attaches to the world of objects much as the artist attaches to his paints, the poet to his or her words. Putting children in a rich object world is essential to giving science a chance. Children will make intimate connections, connections they need to construct on their own. At a time when science education is in crisis, giving science its best chance means guiding children to objects they can love.

At present, there is some evidence that we discourage object passions. Parents and teachers are implicitly putting down both science and scientists when they use phrases such as "boys and their toys," a devaluing commonplace. It discourages both young men and women from expressing their object enthusiasms until they can shape them into polite forms.[35] One of the things that discourages adults from valuing children's object passions is fear that children will become trapped in objects, that they will come to prefer the company of objects to the company of other children. Indeed, when the world of people is too frightening, children may retreat into the safety of what can be predicted and controlled. Many of the papers in this collection recall childhoods at a moment of vulnerability when objects reassured. This clear vocation should not give objects a bad name. We should ally ourselves with what objects offer: they can make children feel safe, valuable, and part of something larger than themselves. These essays demonstrate that objects can become points of entry to larger, transformative experiences of understanding, sociality, and confidence, often at the point of being shared.

In his memoir, *Uncle Tungsten,* neurologist Oliver Sacks describes the importance of old family photographs taken in London during World War II to his developing sense of scientific identity. They provided him, at a vulnerable point in his life, with a sense of stability by giving him objects to catalogue: "They seemed to me like an extension of my own memory and identity, helped to moor me, anchor me in space and time. . . . I pored over old photos, local and historical ones as well as the old family ones, to see where I came from, to see who I was."[36]

The photographs bring more than a sense of identity and history. They are "a model, a microcosm, of science at work." Sacks explains how they provide him with a grasp of a scientific sensibility, and that of a "particularly sweet science, since it brought chemistry and optics and perception together into a single, indivisible unity."[37]

It seems wise to attend to scientists telling the story of their romance with objects. Memoir encourages us to make children comfortable with the idea that falling in love with things is part of what we expect of them. It moves us to introduce the periodic table as poetry and radios as a form of art. Writings that describe the birth of scientific identity make for a deeper appreciation of its nature. Understanding how scientists are made can help us to make more science. Scientific memoir should be part of science education. There, memoir should be written and memoir should be read.

PART I

MIT Students and Their Objects (1979–2007)

What Makes a Scientist?

What We See

MAPS

Steven Schwartz (1982)

During most of my childhood, I had an intense relationship with maps. These were the street and highway maps that were given out for free at gas stations and now cost about two dollars. I got two or three maps each time we filled up the family car; I developed a sizeable collection within a short period.

Each map was a module (with a common interface) in a finite system. With one map for each region in the tri-state area around New York (New York, New Jersey, and Connecticut), I had a well-defined technological system. The highways that linked the regions held the system together.

Of course, for some single regions, such as New York City, I collected many different maps that conveyed the same information in varying depths of detail and in different styles. I often chose which map to use depending on my mood.

Maps use three different language systems to communicate information: street names, compass points, and route numbers. Street names are the most commonly used system, but they lack consistency. Compass points, though very consistent, are not always useful; one must sometimes drive west a few miles to go northward.

Route numbers provide a consistent and useful system. An ordinary person can easily remember a sequence of route numbers. But routes, too, have their own complexity. For me, different levels of routing (interstate, federal, state, and county) illustrate how multiple models of operation can be used independently or together.

I used my maps for three main activities—planning a route, imagining a drive along it, and navigating family trips—all requiring problem solving, fantasy, and responsibility.

Planning a route consisted of translating the question, "How do I get from here to there?" into a line connecting two points and, sometimes, a list of route numbers and street names. The number of points was almost infinite; the number of routes I could design was without bound. By qualifying the problem—"Find a scenic route from Baldwin to Narragansett via Westbury"—I could come up with enough possibilities to keep myself occupied until the federal budget was balanced.

Once I decided on a route, I had a fantasy adventure. Perhaps I had traveled portions of this route before. I could relive old experiences and imagine some new ones. If the highway were one I had never seen, I could make some educated guesses about what to expect along the way. I might build a story around the trip—"Let's go visit Grandma, who lives two interstates and three county roads to the south." The only limitations belonged to the highways—their exits and their speed limits.

The amount of time I spent with maps freed me from physical dependence upon them. I could quote route numbers and street names from memory for cities I had never been near. I took on the role of "family navigator." My parents often asked me for route information, either before or during a trip. By the time I was nine, maps and routing were an integral part of my life. Fantasy and mastery kept me quietly occupied, my mother ecstatic.

The maps offered an infinite number of possible routes, possible fantasies. And they offered me, even when I was very young, a world of mastery and control. I could see optimal routes faster than anyone else, and

I knew map jargon. Few people are familiar with the Interstate Highway System Route code. For example, in a three-digit interstate highway number if the first digit is odd, the highway enters a large city; if it is even, the highway goes around the city. This kind of fluency enabled me to take charge of route planning for my family and also for my parents' friends who would often call with questions. Occasionally, someone would test me with a too difficult question; this was when I used my airport loophole.

I put two "warps" in my system. First, I controlled time. If I didn't want to spend three hours driving up the Atlantic seaboard, I "made" it three hours later and, poof, I had reached Baltimore—without speeding (excessively). If two cities were a great distance apart or I didn't have sufficient maps, I used my second warp: airports. This warp once got me out of an embarrassing situation, when a friend's father asked how to get from Long Island to Los Angeles ("Drive to Kennedy and take TWA").

My old collection of Esso, Mobil, and Hess maps, frayed with age and use, sits in a pair of shoeboxes in one of my parents' closets. I rarely see them, nor do I collect new ones. But I still have a special attachment to maps; they are still a part of me. When I have several unoccupied minutes or am forced to remain in a dull situation for any length of time, I draw highways of various sizes on paper and connect them with an originally designed interchange. If I am feeling ambitious, I draw them in three dimensions, showing their lanes and medians. I always check the finished diagram to verify that the interchange works and all options are available.

Steven Schwartz holds graduate degrees in computer science and chemistry, teaches science in New York, and is currently developing a computational model for how high-school students solve word problems in physics.

PRISMS

Thomas P. Hermitt (1984)

By the time I was thirteen, science was my first love. The physical laws and principles I learned from reading textbooks gave me a sense of satisfying order. Into this world came the prism, and it brought my books to life. I could see the solar spectrum splashed on my wall. I marveled at the simple beauty of nature. I had read biographies of Sir Isaac Newton; now I could see the spectra as he had seen them. Over three hundred years ago, Newton had written: "I procured me a triangular glass prism to try therewith the celebrated phenomenon of colors." Now, I was repeating his experiments in the solitude of my bedroom. Reading Newton had left me drunk with power. A single shaft of sunlight, my trusty prism, and my imagination were all I needed to share in a grand legacy. I felt privileged, part of a proud tradition. I was thrilled by the idea of wondering about the world with every great scientist I had read about and secretly hoped to join the ranks of my heroes. I had not yet taken a high-school physics course, and so I took to my prism and my books, passing hours and cultivating a passion still alive today.

A double window facing south and a large wall mirror in my bedroom seemed to be waiting for me and my beloved chunk of glass. I began with a ritual: I flashed the sun's colors about the room. I especially enjoyed projecting a spectrum down the hall and into the living room, for the size of the spectrum grows with the distance across which it is projected. The brilliance and purity of the colors amazed me. I used several small mirrors and old eyeglass lenses to construct primitive optical systems. I did not strive for large, complex arrangements, but for the simplest ones that would demonstrate the phenomena about which I was reading.

I don't remember when I understood the idea that light was electromagnetic waves propagating through space, but I had this idea as I explored my prism. The idea of waves of very small wavelength refracted according to their frequency turned my curiosity to the problem faced by physicists in the late nineteenth century: what was the medium that supported the wave? Then, the answer was the aether, the hypothetical substance that fills all space and makes the motion of light rays possible. This wispy, ethereal stuff was everywhere, even between the atoms of the glass in my prism. Between them! Visualizing waves of light bouncing off nuclei, slithering through electron clouds, and singing across the vacuum between the stars became an obsession. I never tire of leaving the ordinary, everyday world, shrinking myself down to the size of an electron and diving headfirst into my prism where a front row seat for the spectacle of nature awaits.

Thomas P. Hermitt, who received his SB in physics from MIT in 1984, has worked as a Unix systems administrator at City University of New York Law School and the American Institute of Physics. He is now Senior Unix Systems Administrator for the State University of New York at Stony Brook Hospital and Medical Center.

WALLS

Jennifer Beaudin (2002)

As a child, my favorite pastime was to try to sneak by my mother while she cooked in the kitchen. I tiptoed and darted past her, ducked under the table, into the side room, slinking down low, moving fast. Each time my mother appeared oblivious to me. Then suddenly, she would grab me with a laugh. One day I decided on a more sophisticated approach. I would create a map of the house and a more complex strategy.

I positioned myself in the hallway outside my room and began to sketch. Right away, I had to think about what I was representing. How large was each room? What were its boundaries? I saw that the rooms were made of straight lines. They were all rectangles. As I placed one and then another on my map, I made a startling discovery: a wall of one room could be the wall of another. I had previously imagined the rooms as floating, distinct entities. Indeed, all the rooms were adjacent to each other and formed a whole. I can remember how it felt to suddenly see something new. I walked into each room and tapped on its wall: *this wall is also the wall of the next room.* What was most powerful was the recognition that I had been so familiar with something and yet had not really seen it.

My house invited me to find other new perspectives. I hung upside down off the side of my parents' bed, wondering what it would be like to walk from room to room on the ceiling. I wandered around with my eyes closed, trying to connect this new experience to my prior visual understanding of space. I imagined possible underground connections when I heard the distant voices of family members in other rooms through the vent in the floor.

I made many maps of our house; in one, for a birthday treasure hunt, I had to think through the question: What would my friends not know and not understand about this place? How can they be made to see it? I tried to place myself outside of myself, to empathize with a novice.

I played with my ideas in Lincoln logs, boxes, pillow forts, and drawings. To me, the dollhouse was always more interesting than the dolls. In detailed sketches and cardboard constructions, I experimented with the traditional definition of a house, asking whether it could include a roller coaster or a secret passageway. I tried, without much success, to employ flexible straws for indoor plumbing.

Jennifer Beaudin received her AB in psychology from the Colorado College in 1999 and her SM in Media Arts and Sciences from MIT in 2003. She works as a researcher, specializing in learning and health, with the House_n Research Consortium in the MIT Department of Architecture.

HOLGA CAMERA

Andrew Sempere (2004)

In my final year of high school, my Art 10 instructor arrived late to class, arms loaded with a dusty cardboard box. Dropping the box unceremoniously on our worktable, he held up a handful of black plastic shapes and encouraged us to dig through the box to find a matching set. Parts in hand, he passed around a roll of tape to bind them together. By the end of class, we each had what looked like the fake squirt-gun camera I could never fool anyone with when I was a kid. Our instructor gave us what seemed like surreal instructions: "Make sure to set the exposure to 16 to take 12 pictures . . . you'll know it's right when the number you see through the little window is 13."

Well before this day in Art 10, my father had set up a darkroom for me in our cellar at home. Photographically speaking, conditions there were abysmal. The house I grew up in is a 200-year-old New England colonial that, until shortly before we moved in, had a dirt-floor basement. In that basement there was no running water or air circulation. There was a strange humidity, mold, and spiders. My camera was an old, dirty Pentax with a broken light meter. At first I fell in love with the process, shooting roll after roll, but then I gave up, tired of the discrepancy between the beautiful shots I envisioned and the dirty, ugly prints I was able to produce.

Here I was again. But now my high school's photo lab lacked nothing and was miles away from the basement of the many-legged bugs. Conditions in the Art 10 darkroom were as good as they get. Here I believed only proper equipment stood between me and glory, yet I had just been given a plastic camera to work with. A camera that had to be assembled from parts. Reluctant but curious, I shot a roll of 12, then another.

And so I met the Holga, a camera made in Hong Kong with all the mechanical accuracy and precision of a jar of peanut butter. The back leaked light. The film advance rarely worked. The flash shoe usually malfunctioned. The lens was plastic and distorted the image. On mine, the shutter often stuck. The whole thing obliged one to carry a roll of sturdy black tape, mostly to keep the back from falling off. The Holga was neither more nor less than it seemed—a chunk of plastic that let light onto a piece of paper. By most standards, the Holga was a piece of junk. But as if by magic, the defects in the Holga conspired to lend even the most mundane subjects an air of analog beauty. The lens distortion created effects one cannot repeat intentionally, fuzzy edges and abstract shadows punctuated by dark spikes created by all those pesky light leaks. In spite of myself I fell in love with the Holga because it was so endearingly broken.

As a by-product of loose quality control, every Holga manufactured is different. There is no way of telling exactly what the effect of a given camera will be until one shoots a roll of film. Even with experience, the semi-random nature of the camera means that the photograph you think you are shooting is not likely to be the one that develops later. There is no substitute for doing. Talk is cheap: all that matters is that you make and keep making.

The same year I met the Holga, I took a digital imaging class and learned how to scan and manipulate my photographs. One day my instructor, after listening to my complaint that I didn't have enough of one thing or another to finish a final project, stopped me with a raised hand and the following words: "Look, just build the car with the parts you have, okay?" The message was clear: if I ever wanted my project to see the light of day, I had to learn to work within limits. This phrase codified my feeling about the Holga, reminding me that the defects of the little box are its assets.

The limitations of the Holga are what make it worth using. Accepting that a cheap toy can become a tool for making art means letting go of some myths about art and art making. By the time I pasted together my first Holga, I thought the artist a special, mysteriously enlightened being. I held painting and drawing above other forms of artistic expression. Yet I was conscious that every week, when we lined up our toy-camera photos on the ledge of the blackboard for critique, I saw things I loved. These were images I wanted to keep, ideas worth working on, all created by a team of inexperienced sixteen-year-olds without pencils, paint, complicated equipment or much money.

When I graduated from high school, on my last visit to the photo stock room, I exchanged $10 for a Holga of my own. I never used it. My interest in photography waned, not from dislike but rather as my interest in other means of expression grew. But working in technology, the lessons of the Holga have stayed fresh and served me well. Among these is the notion that to be an artist or a scientist implies a willingness to act as an observer, to keep a record that does not seek to eliminate the blurry edges, dark spikes, and imperfections, but rather celebrates them. This means learning to live with the messiness and disappointments of the real world and working through problems by making things that will in some way fail. Embracing limits, you finally can build the car with the parts you have.

Andrew Sempere received a BFA from the School of the Art Institute of Chicago in 2001 and an SM in Media Arts and Sciences from MIT in 2003. He currently works as Design Researcher for the IBM Collaborative User Experience Group.

What We Sense

SAND CASTLES

Stephen Intille (1992)

The odd thing is that I didn't and still don't like the feeling of sand on my body. I've always found the grittiness uncomfortable, and I don't like the prospect of sand lingering with me after I've left the beach. Today, with fair skin and a dislike for itchy sand, I rarely go to the beach. But even now, when surf-loving friends get me there, I can't help but think how much fun it would be to escape from the "grown-ups" and find a plot of sand for a castle. A big one.

I did not get to construct sand castles often, so when my family planned a week-long summer trip to the shore, I would fantasize about how this year my results would be bigger and better than anything I had built in years past. I envisioned winding moats, high turrets, an entry bridge, roads with tunnels, and high spires.

In some ways, the sand of my childhood castles was like other construction materials I played with as a child. My days were occupied with LEGOs, Erector Sets, and Tinkertoys. These materials made it easy to take a thought and turn it into something real, something you could touch and that others could admire. The construction process was not always smooth. In fact, it always required tinkering with small components of the larger plan. Sometimes an "aha" moment required scrapping the larger plan altogether, because I was missing a building part, a brick or a connector, or did not foresee some tricky situation.

But building with sand was different than working with materials such as LEGOs. Building a sand castle was always a collaborative project. I would start a structure early in the morning, immediately after arriving at the beach. After an hour or two, the growing mass of sand would inevitably attract the attention of my father, my sister, or friends at the beach. At least for a while, others would join in the building process, making moats, wetting sand, or scavenging for shells. Rarely would someone stick around for hours and complete the structure with me, but there was always the feeling of working together. Sometimes strangers would get involved. Near the final stages of a project, after hours of digging and sculpting, younger kids would come around and watch for a few minutes with a parent. I'd describe my progress and my goals, but I would watch them warily, hoping that a sudden bolt didn't take a kid across the top of the castle and leave a wake of devastated sand.

In the confines of my room, working on LEGOs, I never had this sense of collaboration. In school, there was never such a large block of uninterrupted time and a resource as boundless as a good stretch of beach and ocean.

Sand seems light until one starts scooping large amounts from the ground with shells and transporting it around a big space. Walking here, digging down, walking there, smoothing out. Once engaged by the details, I usually forgot about the physical work, until the next day when my arms and middle would be sore from scooping and leaning. I forgot about food, and I got sunburned because I ignored the sun. I used the ocean only to wash down my sand-covered body. There was always the urgency of a sand castle to finish or protect from the dangers at the beach: the wind, the encroaching water.

The hostile elements gave sand-castle building its urgency. Unlike LEGOs or Tinkertoys, castles could not be put aside and worked on hours later without consequence. An awesome tide could seep in and destroy even a structure built high on the beach. A more mundane and predictable threat came from little kids. Many times I left a structure for only a moment, to rinse sand off in the ocean or to apply sun lotion on my burning neck, and would return to find only footprints where a perfect moat and turret had been. Most often, there would be a crying child and an apologetic parent. It didn't matter. I did what I had to do: rebuild.

To build a sand castle requires improvisational skill and a negotiation with the limitations of the sand. In this sense, too, sand castles were a collaboration, but with nature. Sand dries out, making fine detail on small structures difficult to maintain. The solution was to make small details only on large structures. Waves eventually lap at the edge of structures. The solution was to build a protective wall that wrapped around the structure and could hold off the waves for a few extra minutes. Damp sand had a rather uniform color. Enlivenment of the sand became a preoccupation, but I insisted that decoration had to come from the beach itself. Only natural objects could be used (with the exception of dead jellyfish—too disgusting for use in my sculpture). I developed justifications for all construction rules, but all the rules were mine: I avoided using buckets and shovels, but preferred large shells for scooping and hands for molding. Buckets and shovels, well, I used them sometimes, but it was cheating, violating the spirit of the endeavor.

One also had to negotiate with time. It offered so many constraints. When building a moat around a castle, the moat needed to be deep enough to reach the waterline beneath the surface of the sand. However, once that waterline was hit, water would create erosion of the moat walls. Planning was required to ensure that the turrets near the moat were not consumed by the eroding powers of the moat during the construction of the castle. The eroding material needed to be scooped out, but it could be used to make stalagmite-like-castle structures. However, if these structures were built too early, they would dry out before the castle was complete. Timing was crucial. With one mistake, the walls collapsed too fast for repair.

I never read any rules on building sand castles. I came to understand the sand on my own. Each year I would remember lessons from previous years and work on more advanced projects. The sand left me with the desire to make things, motivation for an engineering degree, and a fondness for computer programming. I appreciate nature and simple things built from natural parts. I do not believe that any material thing lasts forever. I learned a certain humility.

The ocean was a brutal, uncompromising timer. The last pleasure of the day was to watch as my creation successfully, if fleetingly, warded off the huge force of the incoming tide. Sometimes I would undertake frantic repairs when surf broke through a seaward wall. Eventually, though, a particularly strong wave would overcome the weakened fort. I roared, part in agony and part in joy. I knew it was coming and couldn't avoid it. I liked that my structure would succumb to the natural force of nature and not the trampling feet of an unknown child. My fear of the castle's destruction turned to pleasure, as wave by wave I watched the sand walls melting into themselves.

Stephen Intille received an SM and a PhD from MIT while doing research on computer vision and computational perception. He is presently Technology Director of the House_n Research Consortium in the MIT Department of Architecture.

THE BODY

Joanna Berzowska (1996)

Because my name was supposed to have been Jake, I felt some pressure to be a tomboy, to climb trees, and carry little transistors in the pockets of my overalls. This did not happen. I grew up uninterested in cars or electronics. When I built things, they were stories, not programmable toys. But I felt a vague discontent with my choices. I liked math and was exceptionally good at it, but it did not fascinate me. I saw the world of technical things and my chosen world as mutually exclusive. I sensed a rupture between the computational and the creative.

Yet my body was a treasure chest of computational gadgets. I counted on my fingers and added toes when necessary. I was interested in how my muscles worked, my lungs functioned, and why my elbow would bend one way but not the other. And I used my body as a model for understanding the world.

So, I understood gravity when I thought about the puzzle of why it took so much effort to hold my arms straight out in front of me for any prolonged amount of time. I remember being six years old and unable to comprehend why it was so difficult even though it involved no apparent work. I eventually understood that the same force that keeps me from floating up into the air also pulls my arms down and that resisting this force requires work. I became interested in how my body moved, by its mechanics. I was confused by the way my Barbie doll moved. I spent many frustrating hours thinking about why her limbs rotated in that awkward and limited way. I compared Barbie's world to mine and composed letters to Mattel laying out options, ways of circumventing the difficulties.

My body was also a solid geometer's object. It gave me my first example of symmetry and taught me about occupying space. In winter, one of my great joys was making patterns in the snow through rotation, reflection, and repetition, describing parabolas with my hands. I used my fingers for addition up to one hundred (I used my left hand for multiples of ten and right for single digits), I measured distances in footsteps, counted the number of breaths between my house and the bus stop, and tried to multiplicatively relate the two.

Joanna Berzowska received an SM in Media Arts and Sciences from MIT for her work in "computational expressionism." She is Assistant Professor of Design and Computation Arts at Concordia University in Montreal. She is Founder and Research Director of XS Labs, where her team develops methods and applications in electronic textiles and responsive garments.

BUBBLES

Matthew Grenby (1996)

I played with simple toys. At three I had a clear plastic bottle, half-filled with a solution of soapy water. I would shake the bottle and thousands of small bubbles would fill the small airspace in the container. Once agitated, any amount of continued shaking would have no effect, except to reshuffle the existing bubbles. And then, somehow, impossibly, the bottle would lighten. The fluid would disappear among the bubbles. If I wanted to play again, I had to wait until the bubbles turned back into fluid. I have no recollection of ever attempting to unscrew the bottle's cap. In spite of my frustration, I was content with the integrity of this little world. The bubbles offered no explicit insight into the ways of their world; rather they left an almost imperceptible impression.

When I turned four, my family moved into a new home, beside which ran a creek. During the summer months and through early autumn I would don my grub clothes and rubber boots, and navigate the path of fallen cedars and ancient, shattered stumps down to the creek bed. Sometimes I would bring a pail with me, perhaps my mother's second-best trowel, but more often than not I would arrive at the creek empty handed. My feet would chill as I stood in the water and gathered large handfuls of clay and mud. The boots would fill partially with water as I stooped to deposit the building material on the creek bed. I would build layer upon layer of mud. Finally, a small wall would come to the surface of the water. I had built my dam. When I least expected it, a small part of the dam would wash away, leading quickly to an ever-larger rift. If I was lucky, I could quickly slap a hastily gathered armful of mud and leaves into the offending hole. More often, most of the dam would be destroyed and I would start over. This little cycle of construction and destruction would play itself out until either I tired of the game or the dam held.

If the dam was good, the creek backed up or found another way to flow around the barrier I had placed in its way. The sun began to set and I was called to dinner, so I went. When the dam stood, my modest achievements blinded me to the larger reality that there was no stopping the creek.

The lesson of the creek was a lesson of small pleasures in the face of indifferent nature. Time after time, I returned to the creek, ready to play, but sometimes the river was only a trickle, other times it raged: in my eyes it simply did not want to play.

When I was sixteen, my high school organized a weeklong camping trip. Our group mapped our ascent of King's Peak, a modest mountain. On the first day of our climb we heard a rumble and then saw an avalanche approach. It stopped about fifty feet ahead of us. The next morning we continued our ascent, ice picks in hand. It was a magnificent day—high clouds, dark blue sky—but I was overwhelmed by fear, an animal fear. My rational self protested, but quietly. I had an acute awareness of the insignificance of my scale in the face of nature.

The lessons: a soap bottle's stubborn complexity, its beautiful disorder; the creek's reminder that we can be blinded by our constructions when we attempt to tame nature; a humility born of the violence of the mountain. It is easy to lose sight of these lessons, cradled as I am by technology and the man-made city that affords me a sense of peace, surety, of my place within the without.

Matthew Grenby received an SM in Media Arts and Sciences from MIT, working in the Aesthetics and Computation Group at the Media Lab; worked for Intel Research; and cofounded iomoi.com, a gifts and e-stationery company he currently runs.

CLAY

Jonah Peretti (1999)

The bell rings. For the next seven hours I will be trapped, alone in a room full of strange children who pass the time transfixed by incomprehensible symbols. I feel invisible and insignificant. I am in third grade but still can't read.

Finally, Saturday arrives, and a big bag of clay is always waiting for me. I grab a string and use it to cut myself a large cube of clay. Then I start building. My favorite technique is slab construction. I roll large flat slabs of clay and bend them to create three-dimensional structures that are almost as big as I am. My specialty is monsters of all kinds. These monsters are much less frightening than the real-life monsters that I face—dyslexia and divorce—and in an effort to escape my private demons I become prolific with clay, producing a new creature every Saturday. Other children stare enviously as I build larger and more complex creations. By age nine, I had presented my work in a one-man show, appeared on the TV news, and was featured in the local newspaper. A lump of clay in my hands was all I needed.

Clay is a perfect object for a dyslexic child. I was not able to process written symbols, but clay enabled me to explore ideas. It merged the concrete and the conceptual. To work in clay, I needed to learn some simple physics. I needed to understand that air expands when it is heated and that clay contracts when it loses moisture in the firing process. This meant that solid pieces of clay usually would not survive because they would contract unevenly and contain air bubbles. Hollow structures, however, were weaker. Their structure required careful advance planning. They needed holes through which air could escape during firing. This kind of knowledge allowed me to create complex forms. I made hinged mouths that could open and close. I made a creature supported by legs as thin and long as a pencil. The trick to making such legs depends on knowing some physics. It goes like this: (1) build and fire the legs; (2) build a body; (3) put the legs in the soft clay of the body; (4) fire again so that the clay contracts and grips the legs. Basically, you use the difference in contraction to physically join the parts.

In the classroom, I tried and failed to memorize spelling words that the teacher assigned. But in the studio, I made things that meant something to me and learned a very different lesson about learning. With clay, I discovered important ideas by conducting experiments of my own design. When I played the game of right and wrong at school I always lost. But when I worked in clay, I learned a different lesson. Here the goal of cognition was not to be right, but to make something interesting, provocative, and original. Things had underlying structures that could be discovered by making them. There were lessons in clay. Clay taught me that I did not need to be like everyone else to be intelligent.

Yet, as soon as I learned to read, I stopped working with clay. I no longer needed to produce monsters. But my clay lessons continued to inform my thinking. I became a technology teacher and spent three years encouraging children to create innovative projects. Now, at the MIT Media Lab, my goal is the same as everyone else's: to build something interesting. I am finally out on the playground with the other kids.

Jonah Peretti is a graduate of the MIT Media Lab and Director Emeritus of the Eyebeam OpenLab, a home for artists, engineers, and hackers pioneering open-source creativity. After teaching at NYU and the Parsons School of Design, he became a founding partner of the Huffington Post news site and BuzzFeed.

MUD

Diane Willow (1999)

I filled a cup with mud and turned it upside down.
I remember that I had to balance the cup carefully. I
felt glowing satisfaction when a cake held its shape and
seeping anxiety when granular pieces of earth cascaded
over the board as I raised the container. I refined my
technique—cup to board—through repetition. I practiced techniques of mud packing, learning the difference between merely filling the container with mud and
pressing the moist earth firmly in it, compressing the
mud to completely occupy the cup's interior. There was
also the landing surface to consider where the top of the
cup of mud that would become the bottom of the mud
cake. To maximize the possibility of having a cake stand
on its own, without collapsing, I attended to its future
foundation. For a straight edge I skimmed the top of the
cup with a spoon or a length of twig.

My own movements seemed crucial. I could lift the
cup in one swift motion or jiggle it slightly before raising
it. I remember when I began to notice that I could focus
on the rightness of the release. Sometimes I used the
surface of the board to help shake the mud loose from
the container, a quick percussive flip onto the board.
Other times the mud seemed so fragile that I practiced
lifting the cup without having it hit the edges of the
newly molded form.

As I worked, I liked it best to sit on dry earth next
to a puddle. Sometimes I adapted a nearby rock as an

impromptu stool. At six, I scanned the edges of neighborhood driveways looking for a slick, viscous paste, moist enough to glint the sunlight with its water layer, thick enough to conform to the terrain of a puddle's depression. I had refined my mud vocabulary so that I could differentiate mud in its various physical states and in the diversity of its composition. That summer afternoon, the rainwater having almost completely evaporated, I was searching for "butter mud."

This was a creamy substance, my favorite type of mud. I liked to use it to coat the surfaces of the forms that I made from grittier muck. My family had moved from the apartment where we lived when I was approaching my fourth birthday to one in a neighboring town. This change in geography brought with it new terrain for muddy explorations. This is where I began to learn about clay. My brother and his friends dug what appeared to be an enormous hole. They were digging to China. In the process, I saw the multicolored layers of earth that had been hidden beneath the weeds that netted our wild play space.

One sticky layer was the ocher-yellow color of the "butter mud" that I prized. As the diggers heaped the dirt away from their excavation site, I carried handfuls to my mud-making area. I played with the moist elasticity of the clay. My hands were covered with a substance that I usually scavenged only in smaller quantities. I had the novel pleasure of finding out what it was like to make entire forms from this material, not simply using it as a surface coating. I came to think that it was warmer than other mud, a metaphor drawn from its color and pliable character. But there was also a physical sensation. What began as a contact with sudden coolness soon encased my hands, forming warm earthen gloves that held the heat of the sun.

My knowledge of my medium grew with each encounter. I can remember hand-to-mud sounds. There was the solid pat, pat, patting of particularly moist mud

cakes; there was the suction as I pressed lush mud between my palms, feeling, then hearing the sticky, sucking whoosh of pulling them apart. The shifting scents
of the slippery, finer grained "butter mud" and the dis-
tinctly textured and moist black mud all come back
to me.

I developed a feeling for mud with hand and eye,
body and mind that still informs my observations of the
physical world. I see it as an early and intense investi-
gation of the poetics of material science, motivated by
my desire to make things and the immediate and plea-
surable feelings of playing in the mud. The more expert
I became, the more I developed the capacity to imagine
being able to think like the mud, shaping volumes and
forms from the inside out.

Diane Willow wrote this essay during her three-year
appointment as artist-in-residence at MIT. A graduate
of the MIT Center for Advanced Visual Studies, she
has worked as an artist and researcher at the MIT
Media Lab and is currently an Assistant Professor in
the new media area of Time and Interactivity in the
Department of Art at the University of Minnesota.

What We Model

GUMBY

Lauren Seeley Aguirre (1985)

Around the age of seven or eight I became fascinated by Gumby. Gumby was a viscoelastic children's toy in an unnatural shade of green. Gumby stood about 8″ tall and resembled a person in that he had arms, legs, a torso, and a head. But there the similarity ended. Gumby looked as though a brick had fallen on him and flattened him out. He was uniformly ½″ thick from head to toe, and his limbs were crude, without benefit of joints, toes, or fingers. For hands and feet, Gumby's limbs merely became slightly wider at the ends.

Gumby's head was most unnatural. It was a lopsided square that seemed gently pushed to one side at the top, so that if one looked at Gumby straight on, the top of his head sloped up from right to left. Gumby had vague, expressionless eyes and a fat-lipped mouth.

One of Gumby's charms was his flexibility. The green rubber of his body was pliable so that he could take on a number of pretzel-like positions. A simple wire structure under the rubber allowed Gumby to maintain whatever position, no matter how ludicrous, I forced on him.

My parents didn't believe in buying a lot of toys, and so I inherited a second-hand Gumby who had previously belonged to my older brother and sister. Because he was so old, his rubber had dried out and cracked at stress points, which allowed me to see what lay beneath—a wire stick figure around which the green rubber was shaped in a mold.

I didn't play with Gumby every day, and when I did, it was just the two of us. I shut my door and climbed up the ladder to my window seat, which overlooked the patio below. My greatest pleasure was making Gumby assume acrobatic positions. He didn't speak to me, or I to him. Nonetheless, he understood me and was supposed to obey me.

I enjoyed playing with Gumby because he could do things I couldn't. Like many children at that age I longed to be double-jointed. I spent many hours twisting back my thumbs as far as they could go, or fiercely splaying my hand so that the palm and fingers became convex. My favorite but most difficult trick involved putting both feet behind my head.

Gumby was better at this trick than I, so he spent many hours sitting with both legs behind his head. Sometimes the stress of this and other pretzel-like positions made his underlying wire structure poke through cracks in the rubber. On such occasions Gumby infuriated me. Not only did the wire prick my fingers, but the discolored metal looked ugly poking out from his arms or legs. I preferred not to see this part of Gumby—his underlying structure, the part that made him work. The extruding wire took away some of his magic. It revealed that he was not flexible because of his own inner strength but because the wire mechanism had made him so. So when Gumby allowed this mechanical part of his to show through, I punished him.

I tried to be fair and make Gumby's punishment fit his crime. In winter I wrapped Gumby in blankets and a plastic bag to keep warm and dry, and tossed him out the window into the snow. In summer I put him in a wooden stroller and left him on the patio. Near the end of my obsession with Gumby I forgot about him until the snow melted and revealed him still wrapped in a Nissen's oatmeal bread bag.

I didn't need or even want to know that Gumby had wire under the rubber to make him flexible. My preference for looking at objects or problems as wholes rather than analyzing their parts continues today. Much to the dismay of my father, I try to feel my way through problems rather than analyzing each step. I don't necessarily need to know how something works in order to feel comfortable manipulating it. Instead I try to pick out patterns and begin to experiment.

Gumby attracted me for another reason—he was not a girl's toy. Although up to the age of about seven or eight I dreamt constantly about being a princess, when Gumby came into my life I had decided that I didn't want to be connected with anything stereotypically feminine and worked hard to dissociate myself from femininity. Somehow I had learned the idea that to be feminine was to be categorized as limited and helpless. Dolls belonged to the world of little girls in pink dresses, whereas Gumby was pleasingly gender neutral.

One summer day when I was about fifteen, my family and some friends went hiking on Mount Washington. Every now and then we met other hikers and stopped to chat. My mother kept telling people that the men were hiking all the way to the top but that the women would soon turn back. Although I was hot and sweaty and didn't want to hike to the top, I decided to do so anyway, just so my mother could no longer say that the men were hiking to the top and the women were turning back.

Through high school I was quite capable in the sciences but felt more comfortable with the humanities. Yet I came to MIT—a place that is associated predominantly with men—in part because it is just one more way for me to prove that I am not limited by my sex. I think back to Gumby—a toy whose wires irritated and sometimes hurt but never seemed like a toy for a helpless girl.

Lauren Seeley Aguirre has been working for the science series NOVA on PBS since she graduated from MIT in 1986, most currently as Executive Editor for NOVA's Web site and other new media.

EASY-BAKE OVEN

Michael Murtaugh (1990)

I got the oven when I was only four years old, after explicitly requesting it on my Christmas wish list. Oh, I really did love that oven, its orange plastic exterior built small enough for a child to carry. The device itself was very simple: a cubical baking chamber supported by narrow black plastic legs, with flat slots extending out from each side, allowing the tiny plate-like metal pans to be pushed in on the left and pulled out on the right when the little cakes were done. The oven came with two or three of the little pans and a long plastic rod with a C-shaped end to push the pans into the oven's shining belly. I can remember staring into the interior of the oven as I pushed the batter-filled dishes along metal guides. The baking was done by a light bulb! When Easy-Bake stopped working it was time for Mom or Dad to open it and change its bulb.

I spoke to my mother about the objects that I played with as a child, and of course the Easy-Bake was what we both remembered. But even after that conversation, I thought I would find something more serious than the Easy-Bake, but it kept coming back to my mind. I was always so pleased to make anyone a cake. I did birthdays, bridge games, casual get-togethers, any occasion at which my efforts would be appreciated. I would really get into icing cakes for special events, and for extra special events, I would bake a battery of the medallion shaped cakes to stack into one towering layer cake. Adults seemed so willing to shower praise on the "young chef," complimenting my work and seeming to honestly enjoy the result. Cooking also seemed like a unique hobby for a boy my age, and my sense of being special was reinforced by the praise I received. Baking, for me, was like learning card games. It made me feel that I was part of the adult world.

As a child, cooking seemed magical, a mysterious ritual, full of exotic ingredients, with its own books of explicit and complicated procedures written in a strange vocabulary. Cooking was a force that drew our family together, and by learning to cook by myself I both defied that bond and demystified the magic. I put the magic under my control. (In the years after the Easy-Bake Oven, I had two children's cookbooks that I pored over. I had to make everything in the books.) And cooking gave me some sense of equality with my mother. Her appreciation made me feel special; later, I could help her with "real" cooking. In my mind I wasn't just helping but working with her in a cooperative way.

And finally, cooking was scientific. One had to learn the language of cooking, to distinguish between the abbreviations for teaspoon and tablespoon, and learn what the different fractions of a cup meant. I began to think about which fractions were bigger than others just by the way plastic measuring cups fit into each other. (One-fourth cup is the smallest and fits into one-third cup, and so forth.) In my early teens I became very attracted to computer programming, an ongoing passion. Like cooking, its essence is that one follows rules and performs procedures. And as with cooking, programming gave me the security that as long as I followed directions, I would achieve the desired result. In cooking and programming, I could accomplish something very impressive just by learning to take the right steps.

Michael Murtaugh, who continues to enjoy both coding and baking, works as a freelance programmer and teaches in the Master of Arts program in Media and Design at the Piet Zwart Institute in Rotterdam, the Netherlands. He received an SB from MIT in 1994 and an SM in 1996, both in the Department of Electrical Engineering and Computer Science.

KITCHEN CLOCK

Emmanuel Marcovitch (1996)

I remember a huge room with an overwhelming table and big chairs. It was my kitchen in 1977, and I am four. On the blue wall, near the entrance door, I see the big clock, the clock on which I learned to tell time. In addition to this clock, there is a smaller digital one on the buffet—the kind of clock you receive for free when you subscribe to magazines.

I remember that it was not easy for me to tell time on "classical" clocks, the ones with hands. For instance, when the small hand is on two and the big on ten, you say that it is "ten to two." But when the big hand goes to eleven, you do not say it is "eleven to two." The numbers on the dial do not indicate what the real time is. You have to translate what the clock says with its hands into another language.

I remember that I used to spend afternoons comparing the time given by the two clocks. It was easy to tell the time on the digital clock, but it was the wall clock that was my first math teacher.

The clock, with fifty-nine minutes in an hour, made the numbers under sixty the focus of my counting. I did not know how to count past sixty. I was too focused on the clock numbers to waste my time on the other ones. The clock's emphasis suited me because numbers over sixty posed difficulties for someone like me whose first language is French. In English, each number after sixty is logical. Sixty is six times ten. And logically, seventy, eighty, and ninety are respectively, seven, eight and nine times ten. But in French things are much more complicated. Indeed, everything is more or less logical until sixty (*soixante* is six times ten or *dix*), but after that the system does not move smoothly on its logical track. Sixty is six times ten, but eighty, is *quatre-vingt,* four times twenty, and ninety is *quatre-vingt-dix* or eighty plus ten. These are not logical numbers. They are named as they are because of the evolution of the French language. So in my mind there were the logical, useful numbers under sixty, and the more mysterious numbers—I called them the "curious numbers"—a coding scheme supported by my hours at the clock.

Many people talk about the gap between the word one uses and the mathematical symbol used to represent it. For me, there was another, equally important gap: the gap between the curious numbers that began at sixty and one hundred, the next "famous number" that everybody knows. When people talked about such curious numbers, I was able to translate them, more or less. But it was very difficult for me to formulate these numbers by myself. They were not in the range of numbers I used for thinking. I put a barrier at sixty: the clock, my personal math teacher, never talked to me about higher numbers. . . . As I write this, a question comes to my mind: what if my parents had put a barometer instead of a clock on the kitchen wall? Would I have been able to learn numbers up to one thousand and above?

The clock was my math teacher in more ways than one. It didn't just teach me numbers, but mathematical concepts such as the notion of symmetry. The axis formed by the hands when it is six o'clock is a kind of mirror. Each time has a logical image in the mirror: for instance, eight o'clock is the mirror image of four o'clock. The clock also taught the order of numbers. For me the concepts of greater-than and less-than had a physical explanation; each number, from one to sixty, has a position on the dial. The more you have to "turn" to place the number, the bigger it is. I was able to put each number in its logical place. And this method helped me a lot when I began to learn trigonometry, the easiest part of mathematics for me.

For a long time I used my clock-based ways of thinking to classify numbers. When someone told me his age, I saw him on a dial. For instance, if he was twenty I visualized him as a point positioned on four o'clock. Young people were the ones who were closer to twelve on the right, old ones on the left. According to this structure of my mind, there were four logical steps for me during life: fifteen, thirty, forty-five and sixty years old.

Why do you have to be sixteen to drive a car? Why must you be twenty-one to drink alcohol? These ages were not logical to me . . . they were juridical. Logical ages would have been for me fifteen or thirty, or at least twenty or twenty-five.

What about someone who was more than sixty? A new turn begins. For instance, someone who was seventy-five was on the same position as someone who was fifteen. Without considering the philosophy behind it, I established this "physical law" that was natural for me. Later, I recognized in it the concept of congruence in trigonometry. Without being aware of it, I created a mathematical structure close to the advanced concepts I would study ten or fifteen years later. For many children, things are alive or at least have meanings. Children talk to them and have feelings for them. In my case, the kitchen clock had a privileged status. I used it to think.

If a person in my life—say, my mother—had taught me how to read the time on clocks, I probably would not have built this mathematical structure in my mind. Teaching myself—but with the clock as a companion—was the best way for me to come to understand and love mathematics.

My parents repainted the kitchen in 1980, when I was seven. They took out the clock. Anyway, it was not useful to me anymore since I had its structure in my mind. They painted the walls white, transforming the kitchen into a room for adults.

Emmanuel Marcovitch studied the social uses of the Internet while a visiting scholar at MIT and has worked for Vivendi Universal and the French Government. He is a 2008 graduate of the Ecole Nationale d'Administration (ENA) in Paris and also has degrees from the Institut National des Télécommunications and Université Paris Dauphine.

MY LITTLE PONY

Christine Alvarado (1999)

Despite my aversion to playing with dolls—why should I waste time pretending when there are things in the real world to build and explore?—when I was about eight, I was given one doll that fascinated me. At that time, small toy horses in bright colors, called My Little Ponies, were popular among girls. I had several small plastic Ponies that I used to play make-believe with my friends. But I had one larger, plush My Little Pony, a bright-green stuffed horse with a vivid pink mane and tail that I played with all by myself. I would sit for hours on my own, braiding and rebraiding its tail.

I developed a system for braiding the tail of my Pony that taught me about mathematical concepts— from division to recursion. When I started, I took the hair on the Pony's tail and divided it into three pieces for braiding. Soon I became bored with a single braid. I then divided the tail into nine pieces and made three groups. I braided each group of three until I had three braids, then took these three braids and braided them together. Soon I was up to starting with twenty-seven pieces (nested down to nine braids, then to three and then one) and then on to eighty-one. All the while I was learning about math: I saw that division is the process of taking a large number of things and grouping them into a smaller number of groups. In order to end up with one even braid at the end, I had to be able to divide the initial number evenly by three, then by three, and then by three again, until I ended up with just one braid. I learned that I had to start with specific numbers of pieces in order for the braid to come out evenly. These specific numbers, of course, turned out to be powers of three.

Overall, though, what I liked most about braiding was recursion. The large braid was made up of smaller braids that in turn were also made up of smaller braids, and I pushed this structure as far as I could take it. I once attempted to begin the braiding process with 243 pieces, but because each of these sections consisted of only about five strands of hair, I was forced to give it up.

With braiding on my mind, I began to see recursion everywhere. One night at the dinner table, I was eating cauliflower and I noticed that it had the same recursive structure of my braids. Moreover, the cauliflower seemed to continue to recurse forever. I began to divide the piece of cauliflower on my plate, determined to find the base level, but it split further and further until the pieces were too tiny to hold. My parents gave me a strange glance, and I continued to eat, still fascinated by the underlying structure of my vegetables.

Christine Alvarado received her SM from MIT in 2000 and her PhD in 2004. Currently an Assistant Professor of Computer Science at Harvey Mudd College with a research specialty in computer interfaces, she is actively involved in outreach efforts to increase the number of women in computer science.

FLY ROD

Cameron Marlow (1999)

When I was a young boy, my father introduced me to the sport of fly-fishing. The fly rod is slender, long, and fragile, and demands a more elegant style of use then the standard spinning rod. When I began, I was four feet tall with a nine-foot pole and had little experience with refined motion. Needless to say, I found fly-fishing a daunting task.

The object of fly-casting is to imitate the motion of a fly suspended on or just beneath the surface of the water. The lure, or "fly," is small and lightweight, constructed from buoyant materials. To enable casting distances similar to other heavier rods with heavier lures, weight is added to the line instead of the lure.

Casting a spinning rod is like throwing a ball; the rod is extended behind the head and flicked forward quickly. This technique is ineffective when you are fly-casting because the weight is evenly distributed on a fly rod. This flicking technique would result either in a motionless lure or one propelled at such high speed that it is ripped off the tackle. Most people learning to fly fish have both of these things happen before they manage the continuous motion that the rod requires. The motion of the line provides constant feedback on the quality of action of the caster's hands. After a few years of quiet frustration, I became master of this casting technique.

Once in high school, I looked up at the motion of my fly line and was reminded of drawings I had made in algebra class. The resemblance was startling, and I could not think of another example in nature of those long, continuous, flowing lines. I realized that the motion of my hand had a direct effect on the movement of the line, much in the same way that the input to a function produced a given output. Without any formal understanding of the physics involved, I was able to see the fly rod as representing a function for which I was the input.

Around that time, my algebra class was studying the concepts of domain and range. I had problems with the material because I was unable to visualize the relationship between the two. My teacher had tried to explain the notion of mapping from one space to another by drawing lines between planes, only confusing me more. I could not understand the process of determining domain and range. But now, fly rod in hand, I realized that no matter what the actions of my hands, I could produce only a certain set of motions in the line. They were constrained by the physical properties of the fly rod. I could also imagine having longer and shorter fly rods with different ranges of motion. Conversely, I realized that there were a multitude of motions that could not be produced by my fly rod, no matter how I moved my hands.

From this point on, the fly rod was my metaphor for understanding function in mathematics. I would often conjure the image of my casting to visualize the relationship between the input and output of a function. After studying calculus and elementary mechanics, I understood some of the physical properties that determined the response of my pole. Instead of imagining the pole as a function, I was now able to determine what function it was.

Cameron Marlow joined Yahoo! Research after completing his PhD at the MIT Media Lab. He currently studies social dynamics for Yahoo! Research in New York.

WOOD STOVE

James Patten (1999)

My childhood home was heated by a wood-burning stove. Every winter evening we would gather in the living room and start a fire for the night. The stove fascinated me. While I huddled next to it to keep warm, I also tried to figure out how to use it and what made it work. I didn't realize it at the time, but I was learning about the transmission of energy and its conversion from state to state.

After the fire was burning it took a while for the outside of the stove to become too hot to touch. Once the outside of the stove was hot, it took even longer to notice the change in temperature in the living room. It took longer still for the heat to make its way upstairs so that the bedrooms were comfortable. It seemed strange to me that the stove would take so much time to heat up the house, but would instantly burn my hand if I touched its side. Holding my hand a few inches away was comfortable for a short time, but actually touching the metal of the stove was not possible. The air between the metal and my hand, an insulator, kept me from getting burned the way I would if I touched the stove itself.

Just as the stove was slow to heat up, it took time to cool down. We placed a thermometer on the back of the stove so that we could prevent it from getting hot enough to cause a chimney fire. My mother told me never to let the temperature get above 450 degrees.

I first watched the thermometer until it reached 450 degrees, and then shut the air vent. I was surprised to see that the temperature kept rising. After some initial panic, I realized that it had begun to rise at a slower pace and eventually began to drop. Once it dropped to a reasonable level I reopened the vent and observed the same trend: again it took a few minutes for the temperature to stabilize. The air vent on the front of the stove did not directly control the temperature, as, for example, a dial on a radio controls the volume. Instead there appeared to be some more complex system involved, of which the air vent was only a part.

Several seconds after I changed the position of the air vent, the noise produced by the fire would gradually begin to change. I began to understand that the amount of air available to the fire affected how the fire was burning and the temperature on the outside of the stove, but that none of these changes was immediate. When I was older, these observations helped me to understand how changes in dynamic systems propagate over time. Later, I would see my experience with the stove as offering clues to the process by which the sun heats the earth. It helped make sense of the fact that noon is usually not the hottest time of the day, even though it is then that the sun is most directly overhead.

I expected heat from the stove to move quickly upstairs when we opened the living room doors, but this was not the case. I was puzzled because I had seen distortions in the air above the stove caused by heated air traveling rapidly to the ceiling of the living room. I expected the warm air to travel to the upper floors very fast. Since my parents were not happy that it took so long for the wood stove to heat the rest of the house, they installed a small fan in the wall above the living room doors, just below the ceiling. When this fan was running, the upstairs became comfortable much more quickly. Even when the fan was not powered, the blades would occasionally turn as the hot air flowed by them.

I understood that without the fan the warmest air had been trapped near the top of the ceiling in the living room, above the level of the doorway. The hot air would not travel to the second floor just because it was higher than the first floor. We had to provide some sort of path for it to get to the upper parts of the house.

I began to build fires in the stove. I collected a bunch of twigs and placed them between two large logs in the stove. I placed a third log on top of the other two so that the logs were parallel and formed a triangle. When I lit the kindling the flame would shoot up and get trapped under the log on top of the pile. Usually that log would catch on fire first, even though it was farthest away from the kindling that was burning. I reasoned that the heat was getting trapped under the top log, and this caused it to catch first. I enjoyed experimenting with different log arrangements, but my triangle configuration always seemed to be the best.

Occasionally I would try to light a fire using paper instead of twigs. The paper would ignite instantly, but the fire would usually not catch, even though it had appeared to be burning brightly. Paper burned more quickly than twigs, but it did not burn as long or at as high a temperature as most wood. It seemed to contain less energy. It seemed important that something of lighter weight had a lower amount of energy. Later when I began splitting and hauling wood for the stove, I saw that the heavier woods such as oak also tended to burn hottest and longest.

From a boiling kettle on top of the stove, I learned that two areas of the stove, although quite near each other, could have remarkably different temperatures. I noticed that if I kept pouring water on the same part of the stove over and over again, eventually the water would stop boiling. However, if I poured water on a different part of the stove, it would still boil. It surprised me at first that boiling water on the stove took energy away from the metal. I used the boiling water to test the temperature of different parts of the stove and began to see how heat traveled through its surface.

Every week we needed to remove ashes from the stove. These we placed in a bucket by its side. I was surprised that the volume of ashes to be removed was so much smaller than the amount of wood that had been burned in the stove since the last cleaning. After a long time, I figured out that the extra mass had to be going out of the chimney in the form of smoke. Before I realized this, I did not think of smoke as having significant mass, so I didn't think it could account for the mass of the wood that seemed to disappear.

The stove taught me physics without my knowing it; my stove experiments led me to theories and experiences I still draw on in my work with dynamic systems.

James Patten, who received his PhD from the MIT Media Lab in 2006 in the area of Tangible Interfaces, creates interactive works in diverse media with themes including performance and social commentary. His firm, Patten Studio, designs and builds computer interfaces for commercial clients.

SHIRTS

Daniel Kornhauser (2000)

I lived in Mexico until I was six years old, then moved
to France because my father had to work there. I still
remember how I took the news at six. I was very happy
we were leaving for France, convinced that I wouldn't
have to go to school because I was certain that I couldn't
learn a second language. I underestimated my abili-
ties. Once in France, I was sent to school right away
and learned French on the spot. It was a painful process. For those first months I was very lonely because I
couldn't communicate with my schoolmates. Suddenly,
winter arrived and I felt cold. I realized I had taken for
granted the warm weather of Mexico City.

I remember being preoccupied with temperature.
Every night I waited anxiously for the weather fore-
cast. In Mexico City, when the sun shines, it is warm. I
didn't understand how the sun could be shining yet the
weather so cold. When I left the house, I felt like a pris-
oner in my clothes; I dreamed of walking shirtless on a
Mexican beach. I was very happy when I rediscovered
the magnifying lens that my father used for his stamp
and coin collection because I knew somehow that the
loupe could merge rays together. I started to recreate
a Mexican point of sun in the palm of my hands. My
real goal was to have Mexican sun all over my body;
I experimented for hours trying to make the little hot
point wider without losing its heat. I never succeeded.
I understood that there was a measurement and scal-

ing problem. I could not know how much I had to concentrate the French rays to make them Mexican rays. I either had very diffuse rays or rays so concentrated that I burned my hand. And since my goal was to have my whole body engulfed in warm sunlight, I could not even come close to the size of the magnifying glass I really needed. I learned that experiments that seem simple and are not simple at all.

After several weeks the weather got even colder and the problem of staying warm moved from outdoors to indoors. Our house was very cold in the mornings; waking up became a painful experience. I asked my father why the house was getting so cold, and he explained that the heat got turned off at night because electricity was expensive. I began to hate the evil janitor who turned off the heat at night (I later discovered that this was an automated process; the janitor had nothing to do with it). I felt powerless against the short days and cold, rainy mornings. All I could do to protect myself against the cold was to leave my clothes on top of the heater so they would be warm in the morning. But one day my family went shopping for special undershirts to wear in cold weather. They had an impressive name, *damart thermodactyl.* When I wore the shirts I stayed warm. For the first time I felt protected against the cold. But for me, the fact that the shirts kept me warm was not their most impressive property. I believed that the real magic in these shirts was that they produced electricity and thus could solve the morning heat shortage.

I accidentally discovered their secret the very first time I wore my shirt lying in my bed and it started to make green sparks. At first I would only get a spark here and there, but after a while I developed my own sparking technique. First, I tucked myself under my bed covers. Then I would rub my back against the sheets. The friction had to be slow; my motion continuous. When I wanted to release the sparks I would separate the top cover from my body and a huge quantity of speaks would fly all around in the darkness under my sheets. *I had tapped into an undiscovered source of energy.* So on cold and wet mornings when I felt sad going to school, I dreamt of a big machine that would rub two shirts together. I imagined two shirts and not a sheet and a shirt because I figured that two shirts would produce double the electricity. This was my secret idea; I shared most of my ideas with my father, but this one I kept to myself. I considered my shirt electricity more powerful than the ordinary kind because mine made long sparks. There were never long sparks when I connected the TV or my night lamp. I was sure that I had developed a way to generate electricity that would enable us to keep the heat on all night long. Finally, I presented my solution to my dad. I got very sad when he responded by asking me, "With what energy would you move the rubbing machines?"

I had other ideas to fight against the cold. There was a solar oven that I found in my favorite magazine, *Pif gadget*. There was an idea for a dynamo in my bike. What I remember best is how wonderful it was to develop ideas in an intensely personal way, to really feel that you owned them. I made a distinction: stumbling upon the sparking shirts felt like luck. The electric power plant was my idea. And it was made more special because it was not a lucky idea; I had been actively searching for it, actively searching for a way to keep warm and I alone had found it. When I got to high school, I learned that my experiments with the loupe had brought me up against the principle of the conservation of energy and my idea for the power plant had stumbled at the problem of entropy. I remember how good it felt to use those memories to think though these principles in physics class. Failure though it was, the power plant brought me to large questions in science. For me, first among these became a belief that invisible things really existed. Magic was not only to be found in fantastic tales; you could find it everywhere, invisibly surrounding you. When I saw sunlight I knew it not only enabled me to see things but was a source of heat; when I felt the fabric of my shirt I knew that sparks were hiding in it waiting to be released.

Daniel Kornhauser received an SB in Electrical Engineering and an SM in Computer Science from the Autonomous University of Mexico, and an SM in Media Arts and Sciences from MIT. He is now a PhD student in Computer Science at Northwestern University.

VACUUM TUBES

Mara E. Vatz (2003)

Three summers ago I was scavenging the curbside trash in my college town, looking to furnish my first apartment. To my surprise, I came across an engineering marvel—wedged in between a rusted mattress frame and a dreary, beat up couch. I thought at first that it was just a table, but as I pulled it from the pile of rubbish, its top slid open, revealing an unfamiliar mechanical contraption. At that moment, its owner came out of her front door to deposit more trash in the pile. "Go ahead," she said, "You can have that." I glanced again at the mysterious gizmo—a circular plate, a mechanical arm, and two dials. I hesitantly asked her if it was, as it now appeared to be, a record player. "Yeah, it belonged to my grandmother," she explained. It was in excellent condition—a 1950s Magnavox bearing its long outdated ten-year warranty sticker—and worked perfectly. "But," she scoffed, "who listens to records anymore?"

That's a tough question. Record players are the toys of a previous generation, practically obsolete. And yet, the Magnavox took hold of me. I carried that record player home, bought dozens of records that summer, and insisted that every visitor to my apartment behave as though they were impressed by the rich sound emanating from the decades-old technology, whether or not they truly appreciated it. I would explain that CDs and mp3s are inexpensive, durable, and of generally high quality, but that records and my record player were, well, just *better.*

To begin with, a record is analog, not digital. Digital takes a "sample" of sound to deliver an approximation. There is no such thing as a "sample rate" on a record. The grooves on a record are translated seamlessly into electrical current, which in turn drives loudspeakers to produce the delicate changes in air pressure that our ears interpret as music. And my Magnavox had tube amps—vacuum tubes instead of transistor amplifiers. A good amplifier strengthens a signal without altering or distorting it. In the case of a record player, the amplifier takes the relatively weak signal generated by the needle as it navigates through the grooves of the record and turns it into a signal strong enough to power the speakers. When it comes to minimizing distortion and conveying overtones, tubes are better than transistors.

I knew my record player had tubes, because I turned the whole thing upside down to get a peek at the circuitry inside. Most of what I saw was familiar—a mess of wires, some magnets for the speakers, a few capacitors and resistors. And then, there were the tubes. I had never seen vacuum tubes before, but I recognized them immediately. There were four of them: glass cylinders that looked like old light bulbs. I could just barely see there was something inside them, but because tubes get extremely hot (and these had been in service for decades), the glass had burned and was tinted a dark hue.

The basic structure of a vacuum tube is simple: two pieces of metal, separated by nothingness, a vacuum. One of the pieces of metal, the cathode, is heated, which allows more electrons to leave its confines, or "evaporate," than would happen at room temperature. The other piece of metal, the anode, is not heated. The temperature disparity between anode and cathode causes electrons to evaporate off of the heated piece of metal and travel across the vacuum to the unheated metal, but not the other way around. In other words, a vacuum tube lets current flow in only one direction. Transistors perform essentially the same function, only on a far smaller scale. But while vacuum tubes have nothing between the two pieces of metal for an electron to latch onto, transistors are made of nonconducting ceramic material, which isn't as clean as a vacuum. Electrons inevitably find imperfections in the ceramic material and settle in. They create an unwanted charge buildup; eventually the signal being amplified is distorted. But vacuum tubes are big and expensive, transistors are not. The market is now geared toward computers—digital is the future of technology; analog electronics are passé. Only very high-end audio equipment is designed with tubes these days.

So, I understood why it is almost impossible to find vacuum tubes in today's equipment. But what concerned me more was why it was so hard to find information about vacuum tubes. Even though they're not popular commercially, they are still valid circuit elements. More than this, they are conceptually important to electro-acoustics.

Yet when I was an engineering student in the 1990s, vacuum tubes were not part of the curriculum. They were mentioned only when professors reminisced about them, the way they might about old friends—brilliant and wonderful friends—that we, students born too late, would never have the chance to meet. I felt compelled to learn more about them. And even though I paged through every modern electrical engineering, circuit theory, and electronics textbook I could find, I didn't come across a single mention of a vacuum tube. It seemed that the market for consumer electronics had taken control of education; understanding of this scientifically significant object seemed to have vanished along with the object itself. It wasn't until I consulted a textbook from 1937—a full ten years before the transistor was invented—that I found answers.

That textbook belonged to my grandfather. He had handled it roughly and with familiarity; I handled it gingerly, cringing at the sound of the old glue cracking along the seams. My grandfather would probably be surprised to learn that modern engineering textbooks aren't written in prose, as his were; they are written in equations. While the explanations in his book are seated appropriately in historical context, my books are overcrowded with uninformative color pictures and tedious practice problems. My engineering textbooks don't discuss the fundamentals or history of the science they will use. My grandfather probably would not understand that engineering was taught to me as a vocation, not an intellectual pursuit.

I learned more from my grandfather's textbook than I could ever have imagined. The textbook provided information about how the tubes worked, but my journey to the book taught me something about the fragility of knowledge. I had always thought information to be timeless, indestructible. But now I see that knowledge is no stronger than the attention and care we show it. We haven't been very careful.

The textbook at least has found a home in my living room. There it lays among vacuum tubes and vinyl, familiar friends, cherished and protected.

Mara E. Vatz, who received an SM from MIT's graduate program in science writing in 2004, is a freelance writer and teaches high-school math and science.

CHOCOLATE MERINGUE

Selby Cull (2006)

I baked my way into science. Baking comes naturally to me. My mother started me as a toddler, rolling out her recipe for "1-2-3 Cookie Dough." In fourth grade, every student had to build a model of a California Mission. Most used Styrofoam or cardboard. I used gingerbread. I baked every wall, doorway, and belfry in that mission, and decorated the garden with shrubs made of Rice Krispies Treats.

By the time I reached eighth grade, I had baked just about one of everything. My mother's *Joy of Cooking* exhausted, I dug up my grandmother's old *Better Homes & Gardens* and found something completely new. It was airbrushed and elegant and impossibly perfect: the chocolate meringue.

There has never been a finer pie. It was challenging, beautiful, and delightfully tasty, all the things a pie should be. The rewards were high and the path perilous: the meringue could collapse, the gelatinous center curdle, the crust disintegrate. A confectionery catastrophe lurked in every measurement.

I started experimenting. If I melted the butter first, was the chocolate center richer than if I creamed the butter? (Yes.) If I used margarine, was the crust moister? (No.) What happened if I increased the cream of tartar? (Disaster.)

It wasn't science—it couldn't be. Science was that pointless subject that I hated. I would sit through chemistry class, copy equations that meant nothing to me, convert from English units to metric, and wonder: "Who could possibly care about this?" Then I would rush home to my kitchen.

For months on end, I baked nothing but chocolate meringue, experimenting with the recipe, trying to understand how flour, sugar, eggs, butter, and chocolate could become a quivering, delicate pie. My pastries, built on trial and error, hypotheses tested and discarded, were science experiments conducted by a teenager who considered science abstruse and irrelevant.

At sixteen, I had to do a science project on how planets form. Miserable, I started reading. It turns out a planet is more than a big ball of rock. When the sun formed, it left in its wake a wide disk of swirling debris: bits of rock, ranging from the size of dust particles to whole mountain ranges. As these innumerable rocks revolved around the sun, they occasionally collided, sometimes bouncing away, sometimes sticking together. Eventually, large clumps formed, and these attracted more material until they were quite enormous clumps. The clumps grew until there was almost no material left around them. It had been a ball of rock the size of a planet, but it was not yet our planet.

Rocks don't like to be mashed together, and the initial growth of our planet was one prolonged mashing. The energy was so intense that most of the rocks melted upon striking the new Earth, and for several million years, Earth was molten. The dense elements, like iron and nickel, sank to the middle of this lava ball, and the lighter elements, like silicon, floated to the top. When it cooled, the Earth became a light, quartz-rich crust and a dense, iron-rich core, with some transition materials in between.

I knew this, I thought. This was baking. Basic ingredients, heated, separated, and cooled equals planet. To add an atmospheric glaze, add gases from volcanoes and volatile liquids from comets and wait until they react. Then shock them all with bolts of lightning and stand back. Voilà. Organic compounds. How to bake a planet.

Literally overnight, geology made sense to me. I knew how to think about rocks, and what's more, I liked it. A chocolate meringue, for all its delicious challenges, lacked any real point, any larger significance. A planet had that.

Like a pie, each rock is a mixture of ingredients, but much more than their mere sum. A pile of flour, sugar, eggs, butter, and chocolate might be edible, but it is not a chocolate meringue pie. A pile of minerals might be a rock, but it is not *this* rock. The ingredients must be added in a certain order. Some react at only a given temperature and only with certain other minerals. Melting the butter makes the center richer.

On my desk, beside my computer, is a tennis-ball-sized lump of iron. Its irregular, grooved surface makes it look like a ball of lint, just emerged from the dryer—a strange, fuzzy appearance for one of nature's densest metals. It is a meteorite, and I learned how to think about it by making chocolate meringue pie and a hundred other pastries.

The meteorite is mostly iron, with some nickel, troilite, graphite, and maybe some more exotic minerals like schreibersite or cohenite. But that's not the point, and merely reciting these ingredients will no more help you to understand the meteorite than dumping a bunch of flour, sugar, eggs, butter, and chocolate in a bowl will give you a chocolate meringue. The meteorite is the result of a recipe so complex and time-consuming that we still have not fully deciphered it. It is from the core of an asteroid—the heart of a bit of rock that might have become a planet, if given time.

I majored in planetary geology. I started a PhD in fall 2006, studying Mars. Geology for me is far from objective. I love the rocks. The minerals have become my friends, almost my children. I watch them, even when I don't have to; I take note when they behave strangely, observe how their characters alter when baked in groups. I still bake almost incessantly, and, when I do, I think about the startlingly complex physical and chemical processes that occurred to create the ground my oven stands on—all transpiring within the delicious gelatinous center of my chocolate meringue pie.

Selby Cull is currently a graduate student in Earth & Planetary Sciences at Washington University in St. Louis, where she studies Mars and bakes bread and chocolate meringue. She received an SM in Science Writing from MIT in September 2006.

What We Play

CORDS

Walter Novash (1982)

From the age of six until I went away to college, two
cords hung down next to my bed. They were not designed to be entertaining—they opened and closed the
curtains on a nearby window—and yet I spent far more
time playing with them than with all of my other toys
combined. The cords themselves were thick and gold.
Tied to the end of each was a cone-shaped metal weight
covered in gold plastic. As I lay in bed, the weights came
down to just above eye level.

I woke up next to them every day and never tired
of playing with them on long lazy mornings when I was
in no hurry to get out of bed. Sometimes I would just
push them away toward my feet and let them hit the far
curtain and swing back, but usually I played with them
in a special way I will call "twirling," although at the
time, I had no name for it, probably because I never had
the occasion to discuss it with anyone.

Twirling involved grabbing one of the cords a foot
or so above its metal weight, and swinging the weight
around in a horizontal plane so that it began to wrap
itself around its own cord above where I had grabbed
it. The gold weight would keep spinning as the coils of
the cord gradually climbed up the vertical center cord,
until at last the length of cord between my fingers and
the weight had been expended, and the weight met the
center cord and stopped. Next, I would start the weight

swinging in the other direction to unwrap the cord. The most difficult and beautiful part of twirling occurred when the cord became completely unwrapped from itself. At this point, my object was to guide the swinging cord with my index finger, so that it would begin to coil itself once again around its vertical part, but in a direction opposite to that of the last wrapping. It was important that this transition be accomplished as gracefully and as effortlessly as possible.

The art of twirling demanded that the weight swing at just the right speed so that it would coil and uncoil itself around the hanging cord uniformly. The index finger needed to apply less and less force as the weight made its way up the cord.

The challenge of creating a uniform coil contributed to the lure of twirling, as did its pleasing visual display. The slowly spiraling path of the spinning weight was very attractive. Two twirling cords invariably interfered with one another, so I concentrated my efforts on the symmetries of one perfectly twirling cord at a time. The flow from one wrap to the next was soothing, almost tranquilizing. Perfect twirling required some concentration, but most of my twirling was done with an empty mind. Once the cycle of wrap–reverse-unwrap–wrap–reverse-unwrap had begun, the motion of the weight took it just where I wanted it to go. Twirling could be done without thinking, and yet it was interesting enough to keep me from thinking about anything else.

In my early years, it is possible that twirling helped me develop an intuitive grasp of the laws of physics and geometry. By experimenting with different lengths of cord, I explored the concept of radius of curvature. The tendency of the weight to keep moving was a fine example of Newton's laws in action. My observations of the weight's pendular motion probably made this concept easier for me to grasp when I learned the equations describing it. But in the moment, all I sensed was the rhythm of the swinging weight that never failed to make me happy.

Walter Novash received an SB in Mechanical Engineering from MIT in 1983. He currently designs and installs renewable energy systems in Madison, Wisconsin.

DICE

When I was little, I enjoyed solitary play; sometimes I even pretended to be sick to avoid going to another kid's house. Playing alone was relaxing. No one was watching me; I didn't have to talk; there were no unfair rules or uncomfortable situations. And I could be the boss. Alone, I made up games, usually based on the professional sport of the season. Some of these involved inventing an athlete and "playing" through his career by rolling dice or flipping cards.

For example, to simulate a running back's statistics in the National Football League (number of runs, total yards) I made up the following rules:

In a single NFL game, a running back will run anywhere from thirteen to thirty-five times and average about four yards per run. There are sixteen games in a season, and good running back can play as many as twelve or thirteen seasons, barring injury.

When I invented a football running back, I would create an entire career, all from rolling the dice. For every invented season, I would decide how many games he would play and how many runs he would average per game. These numbers would change as he became progressively better (three or four seasons), peaked (three seasons), and then gradually deteriorated (four to five seasons).

For each game, I rolled a six-sided die six or seven times and added up the rolls to determine the number of runs for that game. For each run, I rolled the die two or three times, and looked up the sequence of digits in a table to determine how many yards the running back got on that run. I kept a running tally of total yards with a calculator and wrote everything down at the end of each simulated game.

A simulated game was about fifty to eighty dice rolls. There were as many as sixteen games a season, and ten or twelve seasons in a player's career. So, in nine thousand rolls of the dice, I would create an entire imaginary career.

I had similar (but more complex) rules for baseball pitchers and hitters, basketball players, football quarterbacks, and hockey players. Other times I would invent a whole team and make team- instead of player-oriented rules.

I played these games with passion. I came home from school and ran upstairs to finish the game I had started the day before. And when I played, I played with patience. Certain characters took several hours to develop, but I played all afternoon and evening and never got bored.

I learned very early that I couldn't cheat. Frequently I would develop an emotional bias toward this player or against that team. However, I couldn't act on these biases; I felt empty and angry if I cheated and rolled the dice a few extra times. Only fair play satisfied my passion.

I also noticed that a player's career felt unsatisfying (as a work of art?) if his statistics were too predictable from year to year. It's hard to define "too predictable." In some cases, it meant too consistent (relatively similar statistics season after season), in others it meant too unpredictable (no coherency season after season). In real life I could be satisfied with Steve Garvey hitting .317 with thirteen home runs one year and .297 with thirty-three home runs the next year (apparently the result of a plea from his manager during the off-season to sacrifice his batting average for more home runs), but not if my imaginary hero Alex Kirvero hit forty home runs two years in a row.

While the games got me thinking about fairness and predictability, many thousands of dice rolls made me think about the idea of randomness. At eleven, I decided that there was no such thing as a truly random event. Everything has to have a cause. The unpredictability of a dice roll arose from factors like the original orientation of the dice in the hand, the speed at which they are dropped, the elasticity of the surface onto which they are dropped, and the like. If we could duplicate the conditions for consecutive rolls, the rolls would turn out the same. Although my introduction to quantum mechanics somewhat dampened my belief in this theory, I still see its usefulness.

I stopped my fantasy ball clubs because the careers I produced were increasingly predictable. I had made so many of them that patterns became commonplace. And even more important, I produced a career that struck me as perfect. It was like a light shutting off. By the time I was fifteen, I had given up fantasy ball clubs almost entirely, but the all those hours of rolling dice and recording the results played out later in my studies of science and computing.

Jonah Benton is a software architect and developer who has worked on agent-based computing systems, game theoretic simulations, high-performance Web systems, and software for kids. He received an SB in Psychology from MIT in 1992.

EGG BASKET

Erica Carmel (1992)

I was five years old and it was probably April, because I had an Easter basket full of brightly colored plastic eggs. The basket had a long handle so I was able to swing it around in circles. One wall of my playroom was lined with bookshelves that had drawers as well as shelves. They held my doll and toy collection, most of which I never looked at. At the end of the playroom, across from the shelves, was a set of double doors. When I made inventions, I usually included these doors in my designs, probably because their doorknobs were good anchors onto which one could tie things.

I did an experiment with the egg basket. I took a string (in this case, I think it was an extra-long jump rope) and tied it from the handle of a bookshelf drawer to a doorknob of one of the double doors all the way across the playroom. My idea was to create a gondola, such as the one I had seen at Disneyland on a family vacation. I hung my egg basket from the string and tried to run it down the string. When that worked I went on to transport objects from one side of the room to the other by placing them in the egg basket. Next, I moved the string back and forth, causing the basket to swing. As I watched, the basket got further and further above horizontal. Finally, the basket swung all the way around the circle. But, as if by magic, the eggs did not fall out. I was stunned.

I took the egg-filled basket off the string, deliberately turned the basket upside down, and watched the eggs fall out. But when I put the basket and eggs back on the string and once again swung it around, the eggs remained in the basket. I tried the experiment again and again and always got the same results. When they were on the swinging string, the eggs remained in the basket. Yet when I held the basket upside down, the eggs fell out.

I was sure that I had made a new scientific discovery that was going to make me world famous. I ran to share it with my parents. My father was less excited than I had anticipated. He didn't seem surprised that the eggs remained in the basket. He even had a name for the magical force I had discovered: it was called centripetal force. Nevertheless, my excitement didn't die. My father may have known about the force that made the eggs stay in the basket, but I had discovered it on my own. The discovery was mine.

At five years old, I had never heard of the scientific method, but I had followed it. I saw a problem: the eggs remained in the basket when it was swung on the string but fell out when the basket was turned upside down. I created an hypothesis: whatever was making the eggs stay in the basket was only present in the spinning basket. I devised a way to test the hypothesis: I guessed that the faster I turned the basket, the more likely it would be that the eggs would remain in the basket. So, for my experiment, I went back and forth between spinning the basket on the string and then turning it upside down slowly and watching the eggs fall. These results confirmed my hypothesis. There was a definite connection between the speed of the rotation and the likelihood that the eggs would remain in the basket. The conclusions I drew were the most exciting of all: that I had discovered a new principle of science and that my hypothesis was correct. Something "held" the eggs to the basket.

Thirteen years later, as I sat in an MIT lecture hall for my Monday morning class, 8.01, I watched Professor Walter Lewin demonstrate the experiment that I had performed in my playroom with plastic Easter eggs and a straw basket. Lewin took a pail of water and swung it above his head on a string. Sure enough, the water remained in the pail, and Professor Lewin remained dry. At five, I didn't know that centripetal acceleration equals the quotient of the velocity squared over the radius. I also didn't know that for the object not to fall the centripetal acceleration had to be greater than the forces on the object by gravity. What I did know was that the eggs wouldn't fall out of the basket and, as much as the equations are useful, in the end that is all they tell us.

Erica Carmel worked as a management consultant and joined a technology start up in Silicon Valley before going to Harvard Business School. A 1996 recipient of an SM in Electrical Engineering and Computer Science from MIT, she currently works at IBM, managing a team focused on improving customer experience with software.

KEYS

Eric Choi (1992)

When I was five, I often complained of boredom to my mother. Toys for five-year-olds did not interest me. I would play with a toy for one day and put it aside. My mother, exasperated, declared an end to the toys, and gave me a large key ring to play with. No buttons to push, no levers to squeeze, not even a light to light up. Just a ring with keys. I threw it under the bed.

After a few days, it became clear that my mother was serious. There would be no new toys and my old toys bored me. I picked up the key ring. At first, I threw it around because I liked the sound it made when it landed. My father put an end to that game. Because I couldn't throw it, I looked at it more closely and noticed that the ring was not solid as I had assumed. It was a "double" ring. I asked about the double ring and received a curt response: "Figure it out."

Now I saw that the ring was actually one continuous piece of metal wound two times. I could see its end and could pull the two rings apart, enabling me to remove and replace keys. I spent hours taking all the keys off and then putting them back on. I rearranged the order of the keys, sometimes ordering them by how "cool" they were, sometimes by how long they were.

After some time, I thought about opening something with these keys. I started out methodically, trying each key, one by one, on every lock I could find in the house. Most of the time, the key wouldn't fit; when a key did fit a lock, I was delighted but then disappointed when the key refused to turn. I went through every key and every lock in the house. Frustrated, I checked locks outside the house. I tried the family cars. I took the ring to school with me and tried all the locks in kindergarten. I carried the ring of keys everywhere.

I exhausted every possibility I knew, but I was more determined than ever to open a lock. That day finally came when I discovered an old trunk in the basement. By then I knew about different keyhole shapes and which keys were most likely to fit them. This keyhole was different; it was quite simple and unlike any of the others I had seen on doors. One key on the ring also looked unique in its simplicity. I pushed the key into the lock and started to turn. The sound of that lock clicking open was the best sound I had ever heard. The trunk opened. It was empty, but the feat was no less satisfying.

From that point on my confidence grew. I opened a suitcase and a box with a padlock on it. From my point of view, the box contained uninteresting documents, but my mother seemed happy that I had found them; they were something important in the adult world and deemed none of my concern. But this discovery boosted my confidence. I felt useful. My mother had taken notice of me and my new talents. She appreciated my junior locksmithing. And I had been able do something myself, rather than asking an adult to do it for me. Moreover, I was capable of doing something that the adults around me couldn't do.

Soon, I had gone through every lock in the house. I was intrigued by the locks that accepted keys but refused to turn. There must be some way to make them work. I turned my attention to the ridges on the keys, different from key to key, and to a long valley on the sides of the keys that seemed to connect to the horizontal spikes I saw in most keyholes. The meaning of the ridges eluded me since they disappeared when a key went into a keyhole.

I found a flashlight and peered into a keyhole. I couldn't see much, but I did see a little bump close to the front. How can the key go in if that little piece of metal is blocking the way? I managed to use a wire paper clip to push that little piece of metal away. By doing so, I discovered another one behind it. I assumed that this piece of metal, too, must be able to move. I was satisfied that I had figured out how locks work.

My father, having noticed my fascination with locks, showed me a diagram of the inside of a lock from our encyclopedia. My notion of the lock's inner workings had not been entirely accurate, but I was close. I was proud of what I had figured out on my own.

My early lock picking prepared me for working with radios and then with computers, which are much like locks in that they require just the right combination of factors to make them work and teach that one should never be intimidated by complexity or by getting things partially right. Computers, like locks, require that you approach them coolly, with the state of mind of a problem-solver.

If I had taken lock picking to the next level, I might have started to fabricate or modify keys to open other locks, but a law-abiding five-year-old can only do so much. That ring of keys opened more than a trunk or box or suitcase; it opened my mind to a world where many things are locked and the keys aren't made of metal.

Eric Choi, who received an SB from MIT in 1996, held management and business development positions with two Internet ventures, one of them his own. Eric currently is Vice President of a large global financial services firm specializing in the strategic integration and management of elements within the firm's global technology infrastructure.

CARDBOARD BOXES

Janet Licini Connors (1992)

The refrigerator box seemed large because I was very little at the time.

We lived two blocks away from a big department store, and my brother and sister and I used to drive down to the back of the store with my father and take the empty cardboard boxes they had tossed in a pile. The three of us did not care whether the box had once held a stove, a dishwasher, or a refrigerator. We could play with any size. Getting it home from the store to our yard was just the first thrill. Being the one who got to stay in the back of the truck and hold on to the flattened box so it would not fly out of the car was the best! You felt so important—the life of this precious thing was in your hands.

Once we were back home, my parents always made sure that all of the staples or pins were out of the box, and then we were on our own. Many times we played with the boxes as houses. We would cut doors and windows in the boxes; sometimes we connected two to make a house and garage or a living room and dining room. We used them as goal boxes for our games of hockey, as bases for our kickball and Wiffleball games, and as "home base" areas for hide-and-seek.

The activity I remember most, though, was box racing. It started when we took the flaps off a box and put the box on its side. A couple of us would crawl in the box and try to stand it up with us inside. Not knowing much about forces and gravity, the box got pretty bent out of shape and we were quite jumbled up ourselves. We tried to move, but despite pushing in all directions, we did not get anywhere until we figured out that we all had to push on the same side at once. Of course, when we finally tried this, we had too much momentum. Our efforts found us lying on our backs, on our sides, folded over ourselves, and falling out of the box. But when we finally got the hang of standing the box up, we became bored. There is not much to do when you are stuck standing in a box with three other people, besides falling over or getting out. So our next game was to see who could tip the box over first and get their side down.

Once in the down position, we realized that we could roll the box on its side across the lawn. This was fun but painful. We could be gentler on ourselves if only one of us was in a box at a time. One person could build up speed by throwing their body against each side in turn. And things worked better if you crawled instead of lying down inside the box. Pushing on each side with hands and knees made the box roll with a more continuous motion. After many tries, we came to the conclusion that this crawling technique was the most efficient way to roll the box. It became our standard racing position. Looking back, I see that we were applying a scientific method. We tried different ways of moving the box, we made mistakes, and we looked at the results, which we measured in "box travel distance." This is my first memory of learning through physical action. Everything I know and understand I have learned this way.

Now, inside the boxes, and in racing position, we were ready for competition. Somewhere along the way, the other flaps of the box disappeared and what remained was a cardboard tube. On one particular day, one of the older kids was ready in his dishwasher box and I was in mine. Somebody said "go" and we were off, both of us crawling across the lawn in our boxes, our friends cheering us on. My opponent was bigger and he got there first. As the afternoon went on, with different kids paired up, we began to notice that the bigger kids were always able to move the box faster. They kept winning until one of us little guys (the five-year-olds) tried running inside the box instead of crawling. Since we were short enough we could stand and push with our hands while running with our feet. Now, the winning was more evenly distributed across the little kids and the big kids.

The races continued day after day. We learned that cardboard is not especially strong or durable. The more we raced, the more bruised and bent the boxes became. They had no corners left and looked not like boxes but like large telescopes. But to our excitement, they raced even better now. We could roll them more easily because they were not rectangles anymore. It was almost as though we had rediscovered the wheel. As we became more experienced racers, we took the end flaps off right away and bent the corners to create round, sleek racing tubes.

When we raced the biggest boxes, we saw that little kids couldn't move a box, so we turned the competition into pairs racing. Two kids to a box, one little and one big. The pairs racing worked fine except for a steering problem. One member of the team was always stronger than the other and this meant that the tube moved toward the weaker one's side. In pairs racing, the boxes always went off course. One solution was to let only kids of the same size race against each other. More usually, we just allowed the vehicles to go where they wanted. We were all over the yard, into the alley, bumping into each other and into other yard objects as well. We figured out how the boxes worked by playing with them, putting our hands and bodies and minds into the game. The boxes taught me ideas from mathematics and physics and taught me to learn from experience, eliminating problems step by step. Even now, I feel that I know only things that I have experienced and worked through on my own, the qualities we needed to get those cardboard boxes rolling.

Janet Licini Connors graduated from MIT in 1992 with an SB in Management Science, and in 1997 she received an MBA and MA in Arts Administration from the University of Cincinnati. She is currently living in the Philippines with her husband and two daughters.

MUSIC BOX

Gil Weinberg (1997)

It was a plastic white box with a poor-quality speaker
and four colored buttons. Pressing on a button caused
the speaker to play a short melody, a different melody
for each button. I loved this toy and thought it was best
played by hitting the buttons in rapid succession. This
would stop whatever melody was playing after only a
few notes and cause the box to jump to another melody,
which would then stop after a few notes, and so on. The
way I played with the white box, a melody rarely had a
chance to be played from start to finish. At first, what
I most enjoyed was hitting the buttons and changing
the sonic environment. But gradually things changed.
I became familiar with the tunes and I even attributed
abstract qualities to them. I think I still remember the
happy tune and the stupid one.

And then the toy broke down.

I still remember that the first time I pushed a button and nothing happened, it made me cry. When my father could not fix the box I cried some more. I never tried
to fix it myself. When it no longer worked, I put it aside.

But something drew me back. The music box
didn't play, but I pressed its buttons and sang the tunes
that I knew by heart. I reconstructed the melodies as
best I could. After each note I had to think really hard
about where to go next. Should I go up or down? Should
I take a small step or a wide one? I felt myself to be the
melody trying to find its way. This was probably the first

time I wrote original music. Perhaps I could have written music without the white box, but hitting its buttons gave me a start.

I recalled this experience some twenty years later while taking composition classes at the university. I had composed a piece of music and was playing it for a relative who didn't seem to care so much for the piece but was amazed by my process of composition. He could not understand how I "invented" melody. He said that when he tried to write music, he ended up with a song he already knew.

I told him about my experience with the white box. I encouraged him to sing a note, any note, and then think of himself as that note, trying to decide where he wanted to go next. He found the task impossible and blamed it on his lack of musical talent. I disagreed. I believe that musicality can be enhanced at any level and that my experience with the broken white music box had brought me to the method of composing by trying to feel a note, by trying to figure out where it "wants" to go. One of the goals of my work with computers during the past several years has been to introduce new musical instruments to children that demonstrate that everyone can create music, in other words to create broken "white box" experiences for others.

Recently I designed a computer-based music toy and gave it to my young son. It has buttons that activate music and provide visual feedback. My son was happy to push the buttons and hear the music. He is a strong baby and after a week of rough play the toy was broken. I thought back to my own broken white music box and where it had taken me. I wondered whether I should fix my son's broken toy.

Gil Weinberg completed his PhD in Media Arts and Sciences in 2003. He is now an Assistant Professor and Director of Music Technology at the Georgia Institute of Technology.

MARBLES

Kwan Hong Lee (1999)

While I was in kindergarten, my elder relatives introduced me to marbles. Each marble was different but could be categorized by size and color. There were regular, medium, and large marbles. The marbles were transparent, white, dark blue, and black. The most common marbles were the transparent ones; the colored marbles were rare and more expensive. This made them, in my mind, more precious and valuable.

The simplest game was "odds or evens." Two players would play against each other, taking turns in guessing whether there were odd or even numbers of marbles hidden under the other's hand. One player would bet a certain number of marbles and make a guess. If the guess was correct, the second player, the dealer, had to give away the same number of marbles that had been bet. If the guess was incorrect, the player had to give up the same number to the dealer. In thinking back on this game, I remember many arguments about whether zero was an odd or even number. Sometime later, my group of friends came to the conclusion that zero was neither odd nor even. Marbles had somehow led us to a conversation about mathematical theory! Similarly, I could not have known about mathematical notions of pairs before playing marbles, but I was better prepared to understand pair theory when I encountered it by visualizing marbles in my mind.

A few times a week I would count my marbles. Then I would categorize them, saving some precious marbles in secret places. I enjoyed their accumulation. Marbles were like money to me. I learned how to save. I would set a limit on the number of marbles I could go out with, usually ten, because I would play until either my competitors or I had lost all our marbles—basic accounting.

I played a second marbles game, a golf-course style game in which one had to hit marbles from one hole to another. While playing this game, I learned that when objects came into contact, force was transferred from one to another. When I tossed marbles with two fingers, I learned that you needed an adequate amount of force to reach a distance. Through this game I became comfortable with the world governed by physics, Although I did not know any real theories at the time, I had to think about trajectory, rolling distance, and friction because the marble would roll differently on different surfaces. Taking all of this into account, I learned through experience and practice, not through calculation.

In a third game, each player put the same number of marbles in a triangle we drew in the ground. The goal was to use your remaining marbles to knock out marbles from inside the triangle. You acquired any marble you were able to knock out of the triangle. If your marble got stuck in the triangle, you lost and had to withdraw from that game. This game taught me complex strategies as well as geometry. There were safe areas close to the triangle. One had to make many decisions during the game—when to pass through the active play area and when to approach, making yourself vulnerable to attack. The play area was like an experimental lab for physics and geometry.

Sometimes I used to simply play around with the marbles. I would roll them from the top of the slide and be amazed how they accelerated as they rolled down.

They bounced really well when dropped on a hard floor, which made me curious. I understood the bouncing of balls filled with air, but could not understand how something as hard as a marble could bounce so well. I would use their spherical shape to think about planets in the solar system. When I looked through transparent marbles I would see different worlds in them.

I am not sure where my marbles are now, but as a child, they were precious to me, like jewelry and money to grown-ups. I do not remember having any notion of charity. I don't think I gave my marbles away to a younger child. I was possessive about them—they were mine and only for my games. With them I could win or lose, and in my competitive mind, losing was disappointment. As I got older, my marbles became more like an antique collection, and I was drawn away from the familiar playground because my family moved to Korea where no one played marbles. The world had moved on to electronic games.

Kwan Hong Lee received an SM in Media Arts and Sciences from MIT in 2001, studying speech interfaces at the Media Lab. In 2008 he received a PhD from the Media Lab in Viral Communications.

JACKS

Robbin Chapman (2000)

When I was four, most of the older kids I knew played with jacks. They were Serious Jacks Players. High social status and privilege was bestowed on Jacks Masters. An unspoken hierarchy governed levels of participation. Being so young, I was relegated to watching the action, which was fine for me because I enjoyed finding random patterns formed by throwing jacks. Often, when the game was over, I would find extra jacks and use them to make buildings, pyramids, and statues. I was especially glad both boys and girls played Jacks. I hated being told I should or shouldn't play a game because it was or wasn't for girls. In my neighborhood, we all wanted to attain the title of Jacks Master. We articulated our basic scientific understanding of the game in Jacks language: "you have to toss the ball high and hard to gain enough time to pick up all the jacks." But, in the end, a quick mind and nerves of steel are what differentiates a master from an ordinary player.

Jacks reminded me of stars, and I imagined the ball as an asteroid, disrupting their space. Jacks are composed of three lines centered and joined along the x-, y-, and z-axes at right angles, forming a central star shape. Two of the intersecting planes have rounded tips and the remaining have prong tips. Although I didn't possess the vocabulary to describe the jacks' beauty, I found myself drawn to their angles and planes, and the jacks' ability to occupy each other's free space. Either randomly throwing the jacks or purposefully building with them could lead to beautiful designs in space. Unlike some children, I wouldn't play Jacks with alternative game pieces; say smooth stones or little plastic objects. I was fascinated by the geometric properties of the jacks.

As I grew older, I worked hard to become a Jacks Master. We played with a small red ball and a set of ten metal jacks. They were cheap and easy to transport. The game begins by throwing jacks out onto a play surface. Then, tossing a small red rubber ball into the air, you pick up the correct number of jacks and, letting the ball bounce once, catch the ball while holding the jack(s). You work your way up from "onesies" (picking up one jack at a time) to "tensies" and back down to "onesies." A turn continues until you miss the ball, miss the jacks, move a jack, or drop a jack you've just picked up. The first player to go from "onesies" to "tensies" and back down to "onesies" wins.

From the start, I could see variations on the game and was eager to test them. In one variation, we played Jacks in teams. In another I experimented with mixing different sizes, colors, and numbers of jacks. Sometimes we'd pick up only small jacks first, then the big ones. Sometimes we'd alternate the size we pick up at each turn or simply pick up any winning combination. Then there were the "Jacks Jive" challenges. Jives were the chants, songs, or rhymes we invented. Many accomplished players fell apart when they had to stay focused on the mechanics of play while making up new jives on the fly.

I loved the soft rasping sound jacks make when you're sweeping them up from the playing surface. You come to recognize that different surfaces make different sounds. Once I've found my play rhythm, the regular thud of the ball makes me feel exhilarated and unstoppable. To me, the highest form of Jacks involves using extreme surfaces to play on (i.e., staircases, carpeting, bedding). This made for the most challenging play and only true Jacks Masters could handle that kind of pressure.

I learned a lot from Jacks. I developed my creativity by designing increasingly complex game variations. I learned to manipulate increasingly complex mathematical and physical relationships. I developed an intuitive understanding of the intricacies of the physical game. Most of all, Jacks left me with the desire to look at my play and discover its possibilities. The habit has lasted through adulthood. My computer science research asks users to reflect on what they are doing, to reflect on their game.

Robbin Chapman received her PhD from MIT in 2006 in Electrical Engineering and Computer Science, completing her dissertation research at the MIT Media Lab. She is currently Manager of Diversity Recruitment at the MIT School of Architecture and Planning.

PACHINKO MACHINE

Douglas Kiang (2000)

When I was six, one of my father's friends repaid a debt by giving our family a shopworn Japanese pachinko machine. I had never seen a pachinko machine before. It was enormous and seemed to stretch to the ceiling. It had a richly ornamented, battered wooden case and a bright plastic bumper. Its face was glass, like a grandfather clock, and inside the glass I could see row upon row of tiny brass pins, irregularly spaced, and shiny little tunnels and levers mounted vertically on a board. At the bottom of the machine was a long horizontal tray. My father produced a velvet bag filled with shiny metal balls and poured them into the top of the machine. I still remember the raining sounds of all of those balls. Then my father reached behind the machine and plugged it in.

The pachinko machine came to life with simultaneous sound, light, and motion. I stood transfixed as my father pulled back a little silver lever and let it go. One of the silver balls shot up the side of the board inside the glass and trickled down, bouncing off all the pegs in an erratic, unpredictable way until it reached the bottom and disappeared somewhere inside the machine. My father shot another ball and then another and another until a steady rain of steel balls fell over the pegs, bouncing from left to right, never ending up where I thought they would, but invariably causing things to happen along the way. Dials would spin, bells would ring, lights would flash. I was in heaven.

From that day, I played pachinko as often as I could. I began to believe that the movement of the balls wasn't random at all, and neither was the placement of the pegs. A skilled pachinko player can vary the amount of tension on the lever to make the ball shoot through the machine with more or less force. This increases the odds of landing a ball in one of the bonus areas, which rewards you with more pachinko balls in your tray. (Of course, if you own a pachinko machine, you can bend one or more pegs ever so slightly with a pair of pliers and influence the balls' progress in your favor.) As I played, I realized that even though any one pachinko ball could conceivably travel from one side of the playing field to the other during its descent, the best way to land a ball in a bonus area was to keep shooting balls over that area. The pins would deflect some of the balls, but most of the balls would end up in the desired area as they reached the bottom of the machine. The movement of the balls was not random, as I had originally thought.

I also realized that the balls fell at very different rates. Some would fall right to the bottom of the playing board without touching many pegs at all, while others would bounce back and forth among different pegs, taking what seemed an endless amount of time to reach the bottom. It was a revelation to me that when balls bounced off more things, they fell more slowly. (I wondered, if I fell out of an airplane and bounced off enough clouds, would they slow me sufficiently to reach the ground safely?)

As I became more experienced with pachinko, I invented my own games. I shot a ball through the machine and would count to ten, a game in which winning meant that I could get the ball to stay in play throughout my count. I learned to be observant and see beyond the obvious: my goal was to find channels to slow my balls down, but if I shot a ball into an area with many pegs it would most likely bounce into an area of fewer pegs. Finding the slow channels was an art.

I spent a long time searching for that special combination of pegs. If the pachinko ball hit them just right, the ball might bounce back and forth between all of them forever, making me the ultimate master of the "count to ten" game. I never found that special spot where the ball would bounce forever.

One fall day, I watched the leaves from a tree fall in an erratic path to the ground. Although some of the leaves drifted further than others, most of them tended to clump together under the same tree. I thought I understood exactly why some of the leaves fell faster than others. There must be gusts of wind that I couldn't see that were acting like pegs, and the leaves were bouncing off of them.

My experiences with pachinko were the start of my fascination with falling and rolling objects and the reason for my early understanding of basic concepts of math and science. After a while, my love affair with pachinko cooled and was replaced by a newfound and sustained love of pinball.

The Museum of Science in Boston has an exhibit to demonstrate the random distribution of elements on a bell curve. Large black balls fall through a board filled with regularly spaced pegs. The balls all start from the same faucet and the movement of each individual ball is seemingly random, yet after a while the balls invariably fall into a bell curve, with the highest concentration of balls directly under the source. My experience with pachinko had made enough of an impression on me that twenty years later, when I first saw that exhibit, I said to myself, "A big pachinko machine!" I thought the same thing when I learned how to use statistics to make meaningful predictions about the outcome of events. Statistics reminded me of my early attempts to hit the right bonus areas by changing the tension on the lever. When I studied aerodynamics in college and learned how turbulence can affect an object's flight path, I remembered my observations of leaves and their erratic, yet purposeful paths to the ground. It was a big disappointment to me that when we studied the idea of perpetual motion my professor made it clear that such a thing was impossible. The part of me that is a pachinko player still refuses to believe him.

Douglas Kiang studied at MIT while working on his doctorate from Harvard University in Technology and Education. He received his PhD in 2001 and is currently in the Math Department at Punahou School in Honolulu, Hawaii, where he teaches computer science and helps teachers to integrate technology into the classroom.

What We Build

BIKES

Chuck Esserman (1979)

I don't ride my bike. Nor does it remain hidden in my basement. Nor do I wish to sell my bike. I adjust it. I modify it. I upgrade my bicycle's components. I get my bike "right."

Getting a bike right requires an appreciation for perfection. Each component must be the best and carefully integrated with interacting parts. The bike must be polished so that it says that it is fast and responsive—that it has style.

The design of a bike lends itself to this purpose. It is simple. Many parts can be changed. The bicycle industry is always introducing new components that work better or are inscribed with intricate designs. Thinner, higher pressure tires reduce rolling resistance. This enables the bike to turn on a dime. New brake designs reduce the stopping distance. Aluminum parts are replaced by titanium, graphite, and boron versions, each alloy lighter and stronger than the one that came before. Components are slotted and drilled out in various patterns, black and gold anodized, giving the bike a custom look. All of these are required to make a bike right.

Getting a bike right requires continual attention. The brakes must be tightened until the brake pads just clear the rim. Spokes must be tuned until the wheel is in perfect alignment. Derailleurs must shift with pinpoint accuracy. I can always get an adjustment to be a bit better if I work hard enough. These microadjustments require not only skill but considerable patience. One has to feel small, very small, differences.

When I get my bike right, I can't ride it. Twitching my finger on the brake lever causes the bike to scream to a stop, sending me over the handlebars. Each bump on the road damages the wheel alignment, causing the rim to rub against the pads, stopping the bicycle. Shifting gears splatters oil and grease onto the polished components. And after a ride, I must overhaul the bike completely. Regrease the bearings. Replace the handlebar tape. Possibly change the brake and gear cables.

Although I can't ride my bike, I can look at it. I can test the components on a bicycle stand. I can sit on the seat and move the bike back and forth. I can clean and polish it and put it to the side. Then I can look at it again. I know that my bike is right.

Chuck Esserman received an honors degree in Electrical Engineering and Computer Science from MIT in 1980, went to Stanford University Graduate School of Business, and is currently Managing Director, CEO, and Cofounder of the equity fund TSG Consumer Partners.

ERECTOR SET

Kwatsi Alibaruho (1990)

When I was three years old, I spent hours with an Erector Set my mother had given me for Christmas. I began by building familiar geometric shapes like squares and triangles. Soon, I moved to everyday objects like tables and chairs. As my creations grew more complex, I had to take the time to fully analyze the objects I was trying to replicate. Triangles and tables posed few problems; things like buildings and boats required more thought, effort, and analysis.

At four, mother bought me LEGO blocks, and for several months they became an obsession. The blocks were wonderful because, like the Erector Set, they allowed me to build. I could make an object only if I began by visualizing it. Again, I began by building simple household objects and progressed to houses and spaceships. Sometimes I spent days on one project.

In time I imagined worlds both concrete and futuristic. I read Buck Rogers books and watched *Star Trek;* my designs drew on their worlds to build my own. I made sleek cars, hovercrafts, robots, and spaceships. As my LEGO sets got bigger, I designed small LEGO cities. I added dimension to the cities by using pieces from my Erector Set to form larger structures. From time to time I used illustrated books for inspiration. At first this took great concentration. As time passed, building became more intuitive. I needed less time. I was now able to experiment, to use my imagination with more freedom. I would imagine a futuristic something—a vehicle, robot, or building—and I constructed it. I thought of my imagination as constructible. By around six, my obsession with building became more general. When I got curious about something, I wanted to build it.

I began with telephones. I approached them in the frame of mind I had developed with my Erector Set: test mastery of structure by building a physical model. My mother, as always, was my accomplice and encouragement. She provided me with three old telephones, a small wrench, and a screwdriver. I built a model telephone with LEGO bricks. It made an excellent addition to my previous projects. Then, I took to taking apart the telephone—something I loved! Seeing all of the wires and screws inside was an incredible high. In its own way, the telephone seemed just like a LEGO set. Its mechanism consisted of screws and levers that were not too dissimilar to their plastic counterparts. I spent hours engaged with my phones. My goal was simple. I wanted to take the phone apart and then put it back together. The most difficult part of this undertaking was creating clear mental images of what the components looked like when they were fully assembled. But this step was crucial: I would rely on these images as models to reconstruct the telephone. My first several attempts were not entirely successful, but I learned from them. I had no shortage of telephones now; they were offered up by friends and relatives. I deconstructed and reconstructed. Finally, I could take a telephone completely apart and put it back together so that it actually worked. I did not know how the components worked, but I began to get a feel for how they fit together.

In all of this, I was learning to understand structures by visualizing their elements. When I got my first bicycle at six, the first thing I wanted to do was take it apart and put it back together. I had never seen a bicycle up close before. In order to disassemble and reassemble I needed to convert images of bicycle parts into a visual image of a fully assembled bike.

Looking at an illustration in the bicycle's instruction manual provided enough direction to begin the project. I began with some adult help at noon on the day I got the bike, but was able to finish the job on my own in the hours between lunch and bedtime. From that point on, assembling bicycles became a favorite activity. Every two years, I would outgrow my bike and some relative would get me another one. I would feel excited, not from the prospect of having a new bike to ride, but from having a new bike to take apart and put back together. The bicycles that I worked on felt like parts of me.

When I was young, I felt that everything I built with my construction sets was a part of me. Not only did I feel that my creations were special but also I felt I knew them inside and out. No one knew the twists and turns of my LEGO cities better than I. If anyone had asked me where I wanted to live when I grew up I would have said that I wanted to live in one of my own cities.

This need to connect through understanding structure followed me as I grew up. When I became serious about music, I was able to use what I knew about electronics to build a guitar. I recently took apart my electric guitar to make some modifications. After I took it apart, I replaced some of its circuitry with new elements I had designed myself. It is not the perfect guitar, but it is a part of me.

These days, when I write a computer program, I still begin my project with a visual mental model. This helps me to see the big picture. I like to think doing the reverse has power as well. Analyzing the concrete can help to order one's thinking. Now, majoring in cybernetics, I want to take real steps toward creating robots to put in my fantasy worlds.

Kwatsi Alibaruho received his SB from MIT in Avionics in 1994. He is currently a Flight Director for the National Aeronautics and Space Administration.

STRAWS

Austina (Vainius) De Bonte (1996)

I spent the summers of my youth at a Lithuanian summer camp in southern Vermont. We spent time playing sports, doing arts and crafts, swimming in the pond, and singing around the campfire; however, unlike typical summer camps, we spoke Lithuanian while doing all of these things. One of the purposes of the camp was to expose Lithuanian-American campers to Lithuanian culture, arts, dance, and song. One of these folk arts was the art of making *siaudinukai* (pronounced: shau-di-nu-kuy)—decorative straw Christmas tree ornaments.

Siaudinukai are three dimensional—traditionally made out of various lengths of hay or straw strung together with thread. These days, paper or plastic drinking straws are commonly used instead of actual straws. Though it requires a fair amount of manual dexterity, the process is pretty simple: you start by threading together a few straws into a loop, tying a knot to secure it. Then, you continue threading together the remaining straws, each time tying off a loop of one or more straws at the appropriate junctions. The end result is a delicate, yet highly regular structure that can be quite beautiful, especially with embellishments such as tiny straw garlands. Though the most basic traditional shapes are pyramids and octahedrons, a large dodecahedron is the base of one of the most spectacular *siaudinukai,* a twenty-pointed star.

Although I originally learned how to make simple *siaudinukai* as a very young child, I didn't really appreciate the art until many years later when we made them at summer camp. Although most of my cabin-mates were not interested in making *siaudinukai* (many found it difficult to manage the thread and straws while making the knots tight enough to yield a rigid structure), I was enthralled with the idea and was untroubled by the mechanics. Instead of spending time on embellishing them as my instructors thought I should, I was most interested in the structures themselves, and found them just as beautiful unadorned. Once introduced to *siaudinukai,* I spent most of my free time for the rest of camp making increasingly complicated structures. By the end of a three-week camp session, I had amassed more than a dozen *siaudinukai,* which I proudly hung from the eaves of the cabin above my bunk bed.

Most of my collection went far beyond the traditional designs—it was too easy to just copy someone else's *siaudinukas.* Instead, I would take one of the traditional designs as a jumping-off point, experimenting with removing a straw here or adding another there. I added onto existing designs to make more complicated, larger, or weirder-looking ones. Sometimes I would try to make a particular shape, such as an oddly shaped satellite (I had aspirations of being an astronaut back then), or my interpretation of what the Russian space ship *Sputnik* looked like. Sometimes I would just start stringing some straws together, looking for ideas; once something took shape, it was easy to find ways to extend or elaborate on it. The goal was always to make a structure that was solid; it would not crumple when pushed on gently, and each straw's position should be fixed with respect to its neighbors. Often I wouldn't be able to tell for sure whether a complicated structure would be solid until putting in the very last piece. To this day, I have something like twenty of my favorite *siaudinukai* hanging from the ceiling in my bedroom at home, and many more in carefully packed boxes, waiting to be used to decorate a tree during the next holiday season.

Even a simple *siaudinukas* calls for many important engineering and mathematical ideas. One of the most important is making a stable structure, one that is rigid, reasonably strong, and structurally complete. At camp, I even invented my own term for this all-important property: *solidness.* I discovered, mostly by example and through trial and error, that I couldn't make a solid structure that wasn't based on triangles. I also found that every link in a *siaudinukas* was vitally important—the structure was often fully collapsible right up until the very last straw was secured. Furthermore, I discovered that this was actually the mark of a good structure—if the *siaudinukas* was rigid before I was done executing my plan, then quite likely I had redundancies in my planned structure that were not only unnecessary but in some cases actually caused the structure to lose its pleasant symmetry along an axis, hang crooked, or put unwanted tension on other straws. I also found that making rigid structures was a lot harder than one might think. At first glance, it would seem that with a pile of straws and a spool of thread, the possibilities are endless, but I found it hard to make structures that didn't boil down to simple variations of the half-dozen traditional shapes.

There is a lot of procedural thinking involved in making *siaudinukai.* First, you need to plan the overall shape of the structure, and figure out how many straws you will need, and of what lengths. Then you need to find a logical place to start, ideally where you can start with a semisolid shape. For instance, it is easier to manage the mechanics of the project if you start off with a triangle instead of quadrilateral, since at least the angles between the straws of the triangle are fixed (the quadrilateral will be floppy until it is reinforced with triangles). Again, it is easier to build onto the structure if you continue using similar heuristics, trying to build by adding on triangles each time. Finally, it is always a good challenge, though certainly not necessary for a successful structure, that the thread end up at exactly the junction you want to use to hang the *siaudinukas* after you put that last straw in place. Then, all you have to do is make a loop and the *siaudinukas* is complete.

There is a fair amount of debugging involved as well, not only in the execution of a design, but more important, in the formation of its plan. Not all of my *siaudinukai* came out as planned, although hardly any of my mistakes ended up ruining the project. More than once, I was able to turn an apparent mistake into the beginning of a new idea for a different kind of structure. Of course, there were times when midway through a construction, it was clear that a design that had worked in my mind's eye didn't seem to be coming together at all. Then it was time to either change plans and make a different structure out of the existing one, or figure out what went wrong and try again.

Siaudinukai taught me spatial geometry. I quickly discovered the importance of symmetry in making stable structures, and my symmetric *siaudinukai* (pyramids, octahedrons, dodecahedrons) just happened to be important and well-studied polyhedra. When I was actively making *siaudinukai,* I could have told you in short order how many edges a dodecahedron has, since that was just the number of straws I needed to make one.

My experiences with straw constructions laid the foundation for abstract ideas. It was easier for me to grasp the idea of symmetry because I had already experimented with it in tangible straw structures. Making *siaudinukai* gave me a chance to use powerful ideas in a real, physical setting, where I built everything at my own pace. I had my own motivations for building and was free to experiment however I liked, without any concept of a "right" or "wrong" answer other than the standards I enforced on myself.

I like to joke that Lithuanians must have been the world's first engineers, as their folk art is so closely tied to mechanical engineering—the building of bridges and geometric structures. When I watch seventy-year-old women, masters of the art, making *siaudinukai,* I wonder whether they have any idea that the structures that they build from straw are the same ones used to build bridges, glass roofs, and domes. Certainly, I learned a lot from my experiences making *siaudinukai.* I gained intuitions about structural stability, geometry, and symmetry. I was able to apply procedural thinking skills to real-world problems, including planning out an attack and debugging the inevitable errors along the way. And I had fun while doing it.

Austina (Vainius) De Bonte worked as a researcher in the Epistemology and Learning Group of the MIT Media Lab, earning SB and MEng degrees in Computer Science in 1998. Currently she is a Group Program Manager at Microsoft, focusing on planning and product strategy for MSN Messenger.

LASERS

Timothy Bickmore (1998)

I was an extremely shy child, which was somewhat unusual given my profession as a circus performer. At the time I was born my parents ran a circus school in San Jose, California. I was born in its front room. By the time I was three my parents were back on the road. Together, we toured with the circus, my mother doing a single trapeze act and my parents together doing Risley, a two-person style of acrobatics in which one person lies down and uses their feet to juggle the other. I made cameo appearances in the Risley act. I didn't do much, but drew lots of oohs and aahs just for being a cute kid.

I saw my first laser light show when I was a teenager—in a city whose name I never bothered to remember. The laser show consisted of using multicolored laser beams to layer abstract, ethereal designs onto the ceiling of the planetarium. The laser pulses to the rhythm of a rock music soundtrack. What magic—an experience that completely enveloped me. The light show made me feel as if I had become the music. I had found my calling.

I bought and studied every book I could find on lasers and lighting effects and took every opportunity to see laser shows. I would always sit as close to the projector as I could to glimpse its inner workings. One only had a few seconds to see what was within—the few seconds at the beginning and end of the show when the projector covers were removed. I dared not ask any direct questions about the projector, lest someone discover my secret ambition.

The traveling circus community is very isolated, a small town of fifty to one hundred people who live and work together, and who just happen to pull up and move a few hundred miles every day or two. The community is formed at the beginning of each season— when new performers join the continuing house acts, staff, and management of a touring show. It disbands at the end of the season, when most show people take their vacation. Everyone in the community works; there are no sick days or personal days, especially if you are a performer. If you can stand up, then you are expected to be running into the ring with a smile when the ringmaster's whistle blows. If not, you face ostracism by the entire community.

Children who tour with a circus generally follow a correspondence school curriculum with their parents as tutors, although some circus shows are large enough to hire a full-time teacher. Most circus children have very little idea of life outside this community and cannot imagine that public school or colleges have any practical value or why anyone would even want to live in just one place day after day.

My father had a side job, building circus rigging. On several of the shows we toured with, he also worked as mechanic, electrician, and all-around handyman. I've always thought that my knack for things mechanical was at least partly due to the many hours I spent helping him. I know that I became an expert at holding a flashlight under his direction.

When I was fifteen, my parents divorced. My family's breakup meant that I no longer performed. I went into concessions, selling cotton candy. By sixteen I had earned enough money to buy my own motor home. I moved out of my mother's trailer and started to think about what I wanted to do with my life. Of course, it would center on lasers. I decided that it was time to buy a laser of my own so that I could build a projector and embark on my new career. I contacted several companies for catalogs and finally settled on a 0.5mW helium neon laser from SpectraPhysics. I sent them an order using a phony business name, since I was sure they wouldn't sell a laser to just anyone.

Weeks went by, with no response. My panic rose: I was sure I had been found out. I had fantasies about the police coming to arrest me for forging a purchase order. Then one day someone from the front office delivered a plain, medium-size box to the candy wagon where I worked. It was from SpectraPhysics. My heart raced as I tore open the box and foam packaging. And there it was: a two-foot long, nine-inch diameter steel case with an electric cord coming out of one end and a small hole in the other. No on-off switch. No controls. Elegance and simplicity.

When I first plugged in the laser, I faced an anticlimax. It didn't seem to work. Then I realized that the laser beam was only barely visible in the light of day. Dejected, I put the laser back into its box and waited for the end of my shift so that I could go home and try it in the dark. That night I discovered that it did indeed work; I could shine the laser clear across the midway and illuminate a dime-sized red spot on the top of the circus tent several hundred feet away.

I spent the next several months building projection equipment. I bought surface-polished mirrors, prisms, glass, and motors. I constructed special mounts that would let the mirrors spin on the shafts of the motors at a dynamically controllable tilt angle. I bounced the laser off two or more mirrors in succession and then onto a screen. This created spirographic patterns that could be modulated in real-time. I experimented with diffraction by passing the laser through every substance that I could think of (Vaseline on slowly rotating glass was one of the best). I made a projector by mounting all of this equipment on a three-foot square wooden frame.

Whenever the circus kids would get together for an evening party or dance I would bring my laser and provide background lighting. I even put on a few solo performances. In my career as a circus performer, I could barely endure the public's scrutiny: I erected an imaginary wall between myself and the audience. The laser provided a mode of performing in which the audience's attention was not focused on me but on an artifact of my construction. It felt safer. And then there was the sheer joy and satisfaction of constructing something that *I* enjoyed. This sense of safety and satisfaction motivated me to pursue a career in which I design and build things. It was a very different path than that of my parents who chose to please people directly through personal performance. Of course, within my peer group, the laser also provided me with a certain amount of social capital. Knowing that I had achieved something beyond the reach of most of my friends gave me self-confidence.

But the laser provided something beyond all of this. I discovered that the pursuit of science—from discovery to application—could be an aesthetic experience. I felt artistically gratified by the process of designing my projector and using it for performance. Having this feeling has stayed important to me no matter what project I have worked on—no matter how sterile and uninspiring it might seem at the outset.

A year after I built my projection box, my mother convinced me to give college a chance. The only use I could see for such an education was to learn more about how to build lasers, so I selected a university with a good engineering department in a city where I had an immobile (that is, not traveling) relative. I soon moved my studies from engineering to computer science and never got to work with lasers, but the passion that the laser had aroused carried me through the five years of college, a decade in industry, and back to school again. I still seek out scientific projects with some aesthetic component that enables me to perform through my creations.

I have long since lost track of my laser; the last I knew it was in a friend's storage room somewhere in Las Vegas. However, I always think about my time with it as intense and joyous. I am still discovering new ways it enlightened my life.

Timothy Bickmore received a PhD in Media Arts and Sciences from MIT. He is currently Assistant Professor at Northeastern University, specializing in the development and study of computational artifacts designed to build and maintain long-term, social-emotional relationships with people.

LEGO PARTICLES

Justin Marble (1982)

The favorite objects of my childhood fell into two classes: those I took apart (and destroyed in the process) and those with which I built. In the first class, the objects I took apart, there were old radios, television sets, record players, typewriters, electric motors, batteries, and clocks. I was given these objects when they were broken, and somehow, it never even occurred to me to try to fix them. I knew what functioning equipment looked like. I wanted to see what made these things work and I hoped to learn by seeing what they looked like inside. My goal was to find simplicity within apparent complexity. I could not really understand how a TV worked by taking it apart, because its electronic circuitry was not visible, but some things do stand out if you take apart enough TVs. The TV is a complex system built from systems of lesser complexity. These systems in turn contain other systems. At the lowest level there are components that cannot be broken apart—atomic components. Thus did I get an understanding of system design principles.

In the course of reducing electronic equipment to components, I made a fascinating discovery: the most general way of describing electronic equipment was also the simplest. A single TV or radio viewed by itself is a very complex instrument, but each contains only about ten different types of components, and these are common to almost every electronic device. By stepping back and looking at the most general case, the rules become much simpler.

The objects I used for building were LEGOs, a miniature cowboy and Indian set, and all the electronics parts I got from tearing apart radios and TVs. These I vested with special powers and combined them with the LEGOs to provide technological support for self-contained societies of my imagination. I built worlds that included housing, transportation, defense, communication, and power generation. I built my societies using the system/subsystem hierarchical principles I learned from taking things apart. At the lowest level there were atomic objects—LEGOs and electronic parts. When I used LEGOs to build, I constrained their uses. I could be creative only if constrained. Different colored LEGOs had different functions. Red was for war-making machines, yellow for domestic appliances, grey for airplanes, white for general purposes, blue for fuel containers, and clear LEGOs for engines. Electronic components had special functions: batteries were power sources, vacuum tubes were radio transmitters, and capacitors with single leads were bombs. I combined these atomic components into small systems that had well defined functions. I had a way of coupling engines with gas tanks that was the same no matter where they were used. My ships all contained three major subsystems—the engines, the bridge, and the defense. These subsystems contained other subsystems—single gun systems, ship control systems, and power systems.

In everything I built, I needed to work within a structured framework of my own creation. When I received a LEGO fire station kit that told you what pieces to use for what purposes, I found it almost useless. I did not want to build the fire station design as envisaged by the LEGO designers, but many of the parts in the kit were so specialized that they were useless for anything else.

I did not identify with any of the characters in my creations, so I could not act out situations with them. Once the mechanisms had been built and everything was in place, there was nothing to do but destroy them—which I did with relish. My destructions and constructions impressed me with two ideas. The first had to do with the power of hierarchical structure, building systems out of subsystems that in turn contain subsystems. It is a tool I now use to control the complexity of large programs. The second, fundamental to scientific thinking, was how to use induction to make a few simple rules that could explain a multitude of instances. My experiments did not help me with grade school math. I got Cs and Ds in math until junior high school. The radios and TVs did lead me to a hobby in electronics and from there to an interest in computers, which brought me here to MIT. As I was in my childhood constructions, I'm still primarily interested in systems theory. I am still creating complex systems (software) out of atomic functional components within the constraints of good system design. I still have trouble with simple math, but I do have a natural aptitude for abstract algebra and physics. I am still taking things apart, recently starting on my mind.

Justin Marble received an SB from MIT in 1982; he currently is a software engineer in the greater Boston area.

LEGO METRICS

Sandie Eltringham (1990)

Ever since I can remember, it has been an unwritten family tradition for me to receive a set of LEGOs for Christmas. My sister also received LEGOs, but she tired of the traditional blocks in primary colors and instead collected space LEGOs. I didn't understand their appeal because you could make only one object with each kit, with only minor possible variations. My box of all-purpose LEGOs had pieces of many colors; the space LEGO pieces were all gray and blue and made for specific purposes. My sister happily made her spaceships and moon vehicles and then would play for hours, driving her men around and completing different missions. I would help her build, and I enjoyed the challenge of seeing who could find the pieces first. But I did not enjoy playing with the finished product.

Our sets were even packaged differently. My traditional LEGO boxes had pretty pictures of simple structures. Space LEGOs came in boxes with pictures of the completed model in front of a lunar background. And Space LEGO models had decals, which imply permanence. Once they are applied, the piece they cover can be used in only one position. My sister's boxes included step-by-step instructions for completing a model. Even then, I thought that this defeated the purpose of a toy. As I saw it, space LEGOs didn't enable my sister to put anything of herself into her creations.

I played with my first set of LEGOs on the shag carpet of my playroom with sunshine streaming through
the high windows. Part of the challenge was taking a
picture, in two dimensions, and with limited clues, expanding it to three dimensions. I began looking at the
shadows cast by the walls of a structure to decide if
I was duplicating it correctly. For a while, until I got
the hang of three dimensions, I would build only the
front of buildings. But since there were doors in the picture and there was obviously something behind them, I
pushed myself to master three dimensions. Eventually,
I learned to build the houses on the front of that first
LEGO box and even modify them to some extent. But it
wasn't until I expanded my set with new LEGO kits that
my imagination was liberated.

The people in my first LEGO set were heads with
jointed arms. The user built the rest of the body. This
created a person who was out of scale and not very life-
like, which made it hard to pretend it could be human.
I could never play with my people as my sister did, because her space people could fit easily inside her ships
and she could play with them as if they were just small
dolls. I think my frustration with my LEGO people was
at the heart of the analytical attitude I developed toward
building. Unlike my sister, I did not immerse myself in
fantasy. I stayed on the outside. Later, I would get a kit
that included several smaller LEGO people who could
fit through any LEGO door, but it was too late. Instead
of playing with these people inside their houses, I used
them solely as a resource to gauge correct dimensions
for furniture and automobiles. I did not build for the
enjoyment of the little people, but instead for the satisfaction of creating a perfect miniature that even I could
be happy in.

As my collection of LEGOs grew, I no longer lumped
all my pieces together in the same box. I began to sort
my LEGOs by color and shape. I used multiple boxes

and each one contained a different kind of piece. If I grouped pieces, it was easier to find pieces and to estimate if I had enough pieces to finish a job. After playing with my LEGOs, I would carefully take everything apart and sort all the pieces so that I could find what I needed the next time I wanted to play. When I was ready to play, I needed to have everything ready for assembly. My sister found this difficult to understand as she was quite happy to throw her finished products into the box they came in. When I was faced with my sister's half-made toys, rather than undoing them, I would give up and go find something else to do. I derived comfort from order in my life. Controlling something, if only my toys, made me feel safer and more secure.

To this day, LEGOs sit on the top shelf of my closet. This past summer I took them down. It had been a while since I felt the smooth sides with tiny LEGOs written across the bumps and heard the loud snap of two pieces correctly put together. The LEGOs seemed garishly bright. But after a while I carefully separated what I had made and sorted each piece by color and shape and put them away again.

Sandie Eltringham graduated with an SB in Chemical Engineering from MIT in 1994 and went on to receive her MBA. She is currently working in strategic planning and new product development at a Boston area biotechnology company.

LEGO PEOPLE

Alan Liu (1990)

I began playing with blocks at four or five, but I got bored. I could build only so many castles or towers. But LEGOs opened my world. I had LEGO space sets, medieval sets, and advanced builder sets. The LEGO people I put in my world had distinct personalities, realistic jobs, and normal lives. The normal lives of my LEGO people demanded realism and performance from my constructions. I was in total control, and I learned the basics of design and construction.

With practice, I could achieve realism with increasing complexity. I added doors, hallways, second and third floors, and backyards to my buildings. I added furniture. I even built a police station for my LEGO people. The next stage was cars that I built from round LEGO pieces as well as wheels and axles in my collection. I reveled in increasing detail. If my LEGO world was not realistic it seemed of less value.

When I received a Space Set I built a moon colony with a base, rockets, land rovers. As some objects seemed too simple for my increasingly refined taste, I dismantled them for their parts and built more complex structures. Most satisfying to me was that each member of the space colony had a personal identity. I had men and women who had marriages and children.

The space base was built next to a medieval world, with a king, queen, prince, two guards, horses, swords, and flags. I set up the scene suggested by the manufacturer: the royal family enjoying a joust. But I twisted time and culture: I dismantled the horses and gave the king a car. As in the space colony, I gave each character personality. The king was a fool and at the end of each scenario I built, his empire fell apart. The queen disliked her silly husband so much that she spent more time with the prince than with her husband. The prince was a smart aleck who liked to play practical jokes on the dimwitted guards. I applied what little knowledge I had of King Arthur and medieval times to the goings-on in the medieval world.

To achieve other levels of fine detail, I declared war between the medieval and space worlds. The much-despised king always lost, but he lost in several different ways. Before each battle, I built a new weapon or vehicle for the king to use. But he was simple-minded and his schemes never went as planned. The spacemen, in contrast, were always well organized and true heroes in battle; the king and his guards fought on pride alone. I felt some sympathy for the heavily outnumbered king, but I had him lose on his own mistakes. As I built the king's new armory, I always embedded some flaw, some weakness. The king never worked around his weakness; the spacemen always exploited it. And I gave the spacemen another advantage. When the king had a secret weapon, it was somehow leaked to the spacemen. In my mind, the characters were fighting in ways that expressed their personality. The dimwitted king, the elegant and clever spacemen.

After a few years, at about nine, I progressed to the advanced builders' sets. Now I was working with gears and steering mechanisms, rods and axles. I could apply what I learned to working on real automobiles. I began to dismantle the LEGO vehicles in order to create my own. I used the cars, gears, steering equipment, and shock-absorbing springs in the cars of my design. I don't know if I was playing any more. It felt like serious designing. I became fluent with ideas of motion and mechanics. I became an engineer who wants to create and manufacture things. In my mind, the materials of my work have changed but very little else from the days I was building in LEGO.

Alan Liu graduated from MIT in 1994 with an SB in Mechanical Engineering.

LEGO LAWS

Andrew Chu (1992)

I met my first LEGO set in Hong Kong when I was in third or fourth grade and living with my grandparents, uncle, aunt, and cousins. The set came in a big box, and had three or four hundred pieces, maybe more. The pieces came in assorted colors, their shapes quite standard: 2×1, 2×2, 3×2 and 4×2. Most of them were rectangular bricks and had the same number of connectors on top and bottom. Some of the bricks had only one row of connectors on top, the other row replaced by a slanted face that when you looked at it sideways, almost had a triangular profile. All of the LEGOs had a soft texture, almost rubbery, certainly not the brittle feel of so many of my plastic toys.

I began by building houses, castles, and helicopters, pretty standard stuff, and then, with one of my schoolmates, I developed a favorite LEGO game: we each built structures and then smashed them against each other to see which one would break up first. Usually we played to what we referred to as "total disintegration."

It did not take long to determine the rule for when a LEGO structure, a "ship," was dead. For a ship to continue in battle it had to be able to stay on a straight course, with its belly in the water, when it was released. It took us longer to arrive at rules for proper construction of the ships. We both began with ships that looked like a cross between tanks and aircraft carriers, with a bit of spaceship thrown in. These were designs surely influenced by the aesthetics of Japanese robot cartoons.

Our ships were loaded with such features as gun towers, wings, bridges, and gadgets that looked like stabilizers. We quickly learned necessary lessons in structural integrity: things that stick out tend to fall off, and the more things interlock, the stronger the structure. We also learned that when things collide, the smaller the point of impact, the greater the damage. The basic shapes of our ships were reminiscent of the front of an old style train engine, the ram sticking out of the bottom.

Once we determined the proper attack shape for our ships, we worked on stabilizing them. We performed experiments and determined that ships became less stable as they got taller. So, we kept the height of our ships to a minimum. We found square-bottomed ships to be aesthetically unappealing, so our compromise was a triangular wedge, something in the style of a Concord jet. The rear of our ships had to be wide enough to keep them from flipping over when they were hit from the side. Our ships evolved to something close to the shape of the Imperial Empire battle cruisers of *Star Wars,* and we pretty much left it at that.

With construction issues out of the way, we turned our attention to battle strategies. Experimentation was required. We learned that the harder we pushed the bigger the bang, so launching from the floor was best. We discovered that we could get the ships to go faster on a smooth surface than on a rough one, so we used the Formica dining table. However, the size of the table limited our launch distance, so back to the floor we went.

Five years later, I took physics in high school. There, the teacher introduced Newton's Laws of Mechanics and the concept of the center of gravity. This is the point in or near a body at which the gravitational potential energy of the body is equal to that of a single particle of the same mass located at that point and through which the resultant of the gravitational forces on the component particles of the body acts. It affects the stability of a physical body. I realized that I had been experimenting with the center of gravity long before I knew such a thing existed. The extra width I added to the rear of my LEGO ship made it difficult for the center of gravity to "fall outside" of the ship when it was hit. And by keeping the ship low, I learned to lower the center of gravity and enhance the ship's stability.

I had a similar experience of revisiting an old friend when the teacher got to Newton's Laws of Motion. The famous equation $F = MA$ explained why our LEGO ships went faster when pushed harder. One of its derived forms explained why a higher speed was achieved with a longer launch distance. Learning about Newton confirmed other LEGO discoveries about friction, surfaces, and surface areas. The LEGO ships turned out to be a personal laboratory that introduced me to the world of mechanical systems.

About three years ago, as a teenager, I received a LEGO set for my birthday. It was a space system with 278 interlocking pieces. It came with directions for assembling a space base, complete with command center, radar, shuttle craft, and landing vehicle (dune buggy). It included exotic pieces such as mechanical arms with interlocking joints and claws, laser cannons, nubbly tires, steering wheels, little astronauts with helmets and space suits, and rocket engines with fins and lights. If as a child I had been given a set of nubbly tires instead of regular blocks, I might not have ever established the relationship between amount of friction and the number of blocks that come into contact with the floor. I wonder if the new and improved approximations of the real world represented in the space base keep children closer to what already is rather than encouraging them to use their imagination. Besides, the pieces from the new set look kind of brittle. I do not think they will hold up too well under heavy smashing.

Andrew Chu works as a computer network capacity planner for a Boston-based financial services company. He lives in Mendon with his wife and two children.

LEGO PLANNING

Scott Brave (1996)

My love for model building started when I was about
five years old. I did my building with Lego bricks and
what excited me most was following the instructions.
I loved watching how many small and simple steps resulted in a single beautiful and complicated piece. I
found it thrilling that I could take the instructions—
simple pieces of paper—and figure out what they were
telling me to do. This feeling was similar to the one I got
when my sister and I created treasure hunts for each
other. We made clues that led around the house but
always ended with a treasure map. Following the map
was my favorite part. It didn't bother me that my sister,
three and a half years younger, often messed up the
map. I enjoyed the challenge of deducing, from the information on the map, (and what I knew of her), where
she wanted me to go. Finding that place was a thrill.

At eleven my interest in models encouraged my
parents to buy me a kit to build a 4-wheel drive, 30-mile-
per-hour, remote-controlled racing car. Building this car
forced me to go beyond one-step-at-a-time instructions
and taught me to think analytically. I taught myself not
to be constrained by or even to trust what the instructions said to do at any given point. Rather, I learned
to use the instructions and the building materials as
clues to determine my overall path to the finished car:
my treasure.

The model car brought me to analytical thinking
in many ways. To begin with, it was complicated. LEGO

instructions gave me a picture of what I was going to build and made it fairly obvious what to do next. With the racing car, the instruction manual was only the beginning. It couldn't, for example, sort out the confusion of sifting through the car's very similar looking, unlabeled pieces, trying to figure out which piece to use. I learned to disambiguate the selection of pieces before I even began looking for them. Preemptive disambiguation.

In general, to build the racing car, I needed to think beyond what I was told. I needed to work with many different kinds of information in order to make a decision about next steps. I learned to analyze each step before I took it. I learned to plan ahead. And with the racing car, it was important that every decision be correct. Even a small mistake could keep the car from working. And if the car couldn't work, a lot of money would have been wasted. Additionally, LEGO bricks could be taken apart. With the car, building involved gluing parts and threading screws. Decisions were irreversible. LEGO bricks allowed me to figure out the correct thing by experimentation; with the car I had to do the experiments in my mind. After a week or so of work on the car, I would not even begin to decide on a next step until I had looked ahead a few pages in the manual to see how the next piece would be integrated into the design as a whole.

In this way, any thoughts I might have, for example, about step 5b were highly considered, because while there, I had looked ahead not only to steps 5c, d, and e, but to step 6 as well. Sometimes I followed a particular building piece through to its place in the finished car. If I was unsure whether the building piece in my hand was the one intended, I would investigate all parts of its type. I would scan the manual for drawings that looked like my piece and count the number of times it appeared. Then, I would compare my piece to the number still available in the kit to make sure there were at least as many as the design called for. On several occasions, this method saved me from using the wrong piece. If I remembered using the piece before, I would look back to the illustration of that step in the reference manual picture to see if the piece was represented similarly.

These techniques reflect a way of thinking, a systematic planner's way of thinking. The manuals that come with models are written as "assembly instructions." They encourage a narrowing of perspective. The assembly of the model is laid out in self-contained steps that seem doable in isolation. But success is more likely if one sees the process in broader analytic terms.

When I approach any problem today, mathematical, practical, or even social, I remember my experience building the racing car. In my mind, there are always instructions, real or implied, that can guide me in solving a problem. But my best solutions will come if I think with and then beyond the directions.

Scott Brave, an active inventor, received his MS in 1998 from the MIT Media Lab and his PhD from Stanford in 2003 in human-computer interaction. He is Cofounder and Chief Technology Officer at Baynote, Inc.

LEGO REPLICAS

Dana Spiegel (1996)

What did I spend most of my time doing as a child? I took some things apart and I put others things together. So, I would often take apart my GI Joe action figures, removing screws, mixing and matching appendages. Or I would take apart a partially functioning radio to see if I could figure out why it wasn't working. But then, I would sit in front of the fireplace for hours, building large LEGO structures, trying to bring to life a model in my head. These structures were most often replicas of everyday objects. I remember struggling to build a guitar body from small plastic LEGO pieces. I remember another time when I tried to build a telephone with working buttons. I explored the world by trying to re-build it with my LEGOs. To do this, I had to disassemble that object in its real world form, peeling it apart, layer by layer, if only in my mind. From the time I was eight, I taught myself about the world through disassembly and assembly.

I brought LEGOs into my mind; in effect, I cre-ated virtual LEGOs. This childhood practice taught me a great deal: I can now easily look at a structure and recreate it using only a minimal amount of information from schematic diagrams. Looking at a sample product means more to me than diagrams or design specifica-tions. I extend this way of thinking to conceptual prob-lems. I can solve a calculus problem by inspection, by figuring out how it is formed by its constituent parts.

As a computer programmer, I do best with a segmented approach. I break down any problem into smaller chunks, seeing how it is put together as though it were made from basic building blocks. Disassembly. Then I assemble those blocks in code. As it was when I built with LEGOs, the process is easier if I can play with some working version, however rough, of the final product. I recognize my LEGO thinking style when I find the derivative of a complex mathematical equation or build line by careful line of code.

Dana Spiegel, who received an SB in Psychology from MIT in 1999 and an SM in 2001, is Executive Director of NYC Wireless, a nonprofit corporation that helps create free public hot-spots in New York City, and Founder of sociableDESIGN, a company that specializes in the applications of social software.

LEGO CATEGORIES

Joseph "Jofish" Kaye (1998)

For eight years of my childhood, LEGOs were as much
a part of my childhood as my dog Jake, climbable trees,
and a great bunch of packing crates that my sister and
I used to make rooms and endless passageways. But
the LEGOs meant the most to me. And they are with me
today, in how I think and work.

LEGO was always my domain: my sister was never
into it. My parents never played with my LEGOs with me.
I didn't want them to. They wouldn't have understood.

My best friend up the street, Matthew, had Playmobil structures—which I always thought of as in-
herently inferior to LEGO. Playmobil was a series of
preconstructed, snap-together castles and the like. We
occasionally talked about putting his Playmobil together
with my LEGOs, but we never did—the scale would be
all wrong, the Playmobil were people twice as big as
the LEGO people. Incompatible systems. I'd occasionally play with Brio, another Scandinavian product: a
wooden railway setup. But while it was tastefully made
of wood, play was strictly limited by track. There wasn't
too much to create.

To me the difference between Playmobil and
LEGOs is stark: you play with Playmobil; you create
with LEGOs. I never spent much time zooming my
ships around the room or enacting stories or wars
with my creations. I was always building something
better, bigger, cooler. One of my greatest designs was

one I completed at around age seven: a giant glacto-destroyer-mothership-esque thing. The best part of the design was its modularity. It had a superstructure at its center, with a whole lot of launching bays and the like, which were based on spaceships I had already built. I didn't need to smash up all the things I had already built to make the galacto-destroyer.

I remember so much about LEGOs. Like biting the pieces to get them apart. I still use my teeth to strip wires and open packages, and it brings me back to the feeling of working with LEGO. I remember the feel of stepping on LEGOs in the dark, which has to be one of the most painful experiences anyone has ever had. I think about the color schemes themselves. Most of mine were blue and transparent yellow space combinations, but when LEGO came out with black bricks and orange transparent parts a few years later it was astoundingly cool. (Moving along with LEGO styles was my first connection to the idea of culture in flux.)

Every so often, I would reorganize my collection. My most successful organizing system was old cutlery drawers with moveable partitions. I remember anguishing over whether I should organize by color, size, or the number of holes in various orientations. At one point, I figured out what seemed to the right way, the way that LEGOs were meant to be organized. My solution called for different axes for color, length, width, height, and Technics-holes. This was my first foray into thinking in more than three dimensions. I realized then that this was a big idea. This challenge of considering LEGO categories, organization, and storage made it easier for me at a later date to approach programming multidimensional arrays. I remember that when I first met them I consciously used the thought pattern that had worked in LEGO organizing.

I stopped playing with LEGOs when I was eight. I'm not sure why. My family moved from Paris to Singapore, where I was spending more time outside, in the sun, and in the pool. When we moved again, this time to Tokyo, I spent more time exploring computers. I couldn't say if I worked with them or played with them. Those categories made no sense to me. Even now I find that the kids who have the best technical knowledge about a system are the ones who have been given time not just to read the manuals but try all the options.

I donated my LEGO collection to my local elementary school's LEGO/Logo project. The project had only the official parts of the LEGO/Logo kits, and some of the pieces I had accumulated were a lot more interesting to work with. I feel now, and felt at the time, that it was a fitting end for such a collection.

Joseph "Jofish" Kaye received an SB in Brain and Cognitive Sciences from MIT in 1999 and an SM in Media Arts and Sciences from the MIT Media Lab in 2001. He is currently a doctoral student at Cornell University, where he studies information science, with an emphasis on experience-focused computing.

What We Sort

WALLPAPER

Todd Strauss (1982)

So there they were, covering the walls in my bedroom from the time I was five or six until my thirteenth birthday. I do not remember who picked out the pattern, me or my parents. My father hung it.

The pattern was rather plain: simple vertical columns of groups of soldiers, three to a group, on an off-white background (either the color was off-white originally or became that way from dirt and wear). The soldiers were from the Revolutionary era: American, British, French, and Hessian. The groups of three consisted of American, American and French, British, or British and Hessian—never American and British mixed. The soldiers stood at attention, holding muskets, stares fixed straight ahead. Alongside some groups of soldiers was a single figure, in the uniform of the same army, but more distinguished. He brandished a pistol. This was an officer.

I was intimately acquainted with many of the soldiers, some of whom had nicknames. Armed and powerful, they guarded me while I slept, yet I was so much bigger than each of them. I spent most of my time outside the room, so I acted as a messenger, bringing them news of the outside world. Since I moved between the bedroom and what lay beyond, certain things needed to be sorted out. For one thing, there was the question of my size. I had relative size, big in the bedroom, small on the streets, and an actual size, which remained the same. For another, in my room all the action took place in my head, for the soldiers did not talk, but I imagined them speaking. Were my conversations with the soldiers real with respect to the world outside?

There were other questions. The soldiers existed only on flat paper, yet in our interaction they were three-dimensional. Which were they, or were they both? How many worlds are there? I often touched the wallpaper, but to feel flatness was not interesting. There were no edges; there was no movement. Are pictures real? Which is more real: an object (say, an apple), a picture of an object (say, Cézanne's apple), or the picture itself (the brushstrokes and canvas)? What is the relationship between the picture and the object it represents?

Since these representations on my wall were not just objects but historical figures, soldiers of the Revolutionary War, I became dimly aware of the concepts of consciousness and historical remembrance. The war of my wallpaper world was over. I knew that. Did the soldiers know that? But which was fact: the story of the American Revolution as I learned it or the battles I imagined? Are all we know of the past our interpretations of it?

Todd Strauss received an SB and SM in Mathematics at MIT and a PhD in Operations Research from the University of California at Berkeley. For the past twenty years, he has been working on issues related to energy and the environment in the academic and business world.

TOY MAILBOX

Britt Nesheim (1989)

My mailbox was red, blue, and white and stood about 15″ tall, 6″ wide, and 4″ thick. At the top were six vertical slots that stood side by side. The tallest slot was 2″ tall and ½″ thick. The shortest was 1″ tall and ¾″ thick. The slot on the far left was the tallest and thinnest while the slot on the far right was the shortest and widest. It came with six colored disks that varied in diameter, thickness, and color. The tallest and thinnest was red; the yellow disk just a little shorter and wider, and the blue disk was shorter and wider than the yellow disk. This color scheme repeated itself for the other three disks, so that the fattest and shortest disk was blue. Each of the six disks fit one of the six slots at the top of the mailbox.

To play, you inserted a disk into its corresponding slot. Once a disk was placed in the correct slot, it rolled down a plank inside the mailbox, bumped against the back of the mailbox, and slid down two ramps. Somewhere near the end of the path, the disk passed over a lever that rang a bell. At this point, the disk reset against a door at the bottom of the mailbox. The door swung open on two hinges on either side of the bottom of the mailbox so that when you opened the door it formed a little ramp from the mailbox to the floor. If you opened the door, any disk that had been inserted into the mailbox rolled out. All of this I visualized, because my toy mailbox was not transparent. I have no idea what the inside looked like. As I child I could only guess about the mailbox insides by the sounds that the disks made as they passed through.

I became fascinated with the disks and slots. I invented many games for my mailbox. One of my favorites was to leave the door open and put the disks into their slots as fast as I could and watch them come rolling out. I would insert the disks in order according to size from left to right, loving how the disks sped out the door. The best part was that the bell rang over and over as the disks rolled over the last ramp and out the door.

I was a perfectionist, liking things neat and in order. I once thought the mailbox developed this quality in me. But I may have found the mailbox because it let me be me.

It was all so tidy: a large and small red disk, a large and small yellow disk, and a medium and tiny blue disk, each in its own slot. What I loved most was that the medium blue disk belonged in the slot next to the slot for the small red disk. This transition from large to small fascinated me because there was not a big difference between the large blue disk and the small red disk in height and width. In fact, the changes in height and width for all of the disks were proportionate, but the separation between the large and small groups was important to me. I was fascinated by the fact that a large disk could be so close in size to a small disk.

My mailbox trained me to place things into categories. I made two categories of disks: large and small. If I picked up a large disk and it was red, it belonged in the slot to the far left because this was the first slot of the large disks and the first color was red. If I picked up a small disk and it was yellow, then I knew that it belonged in the second slot from the far right because the small disks belonged in the three slots to the far right and yellow was the middle color of the color scheme.

After several years, the plastic that formed the top of the three slots to the far left broke off from overuse.

Once this happened, it was possible to put a disk into a slot in which it did not belong. But not every disk could fit in the three new big holes. The height had changed, but not the width. I was even more fascinated. Now, I could put the large red disk into the slot made for the large yellow disk, but I could not do the reverse. Putting the red disk into the slot for the yellow disk led to a problem, because once the red disk reached the end of its trail, it would not come out. The exit of the mailbox had the same structure as its top. The red disk was blocked. I worked around this by getting the disk to go backwards, back to its red slot. I had to tip the mailbox so that the disk would roll out. Since I could not see the path it had taken, I tried to imagine the inside of the mailbox from the sounds that the disk made when it rolled down. I used these sound clues to tip the mailbox to get the disk out.

As a child I had to think about what was behind the opaque walls of my mailbox. I think of the mailbox when I solve math problems or figure out a complex model. I learned to be a scientist by putting things in categories and listening for sounds that gave clues about structure.

Britt Nesheim worked as a research assistant on the early analysis of MIT papers on evocative objects. She attended MIT from 1989 to 1991.

STOP SIGNS

Joseph Calzaretta (1992)

By the age of two, I could recognize certain shapes as letters and identify them by name. Not long after I read the letters on the red sign at the end of my block: STOP. When I asked my parents about the sign, they told me it was a stop sign and that people had to stop for it. They pointed to a moving car and told me to watch the car's actions. The car came to the sign, slowed to a halt, and then turned the corner. My parents had told the truth.

I fell in love with the stop sign. Every time we passed one on foot I would stop for a few seconds. I would point them out in the car and was delighted when we stopped, respecting the sign's wishes. I owned a picture book and I would always turn to the page with the stop sign and cry, "Stop!" Noticing my fascination with the sign, my parents bought me a stop sign piggy bank. My aunt knitted me a stop sign rug and my father eventually gave me a real stop sign that had fallen off its pole after a car accident.

After the stop sign taught me to read, I discovered letters and words everywhere. But signs had words that commanded people.

I couldn't understand why anyone would ever purposely disobey signs, although I saw that my fellow children sometimes pretended to fool signs by pretending not to see them. As for me, for a while I was obsessed with following the rules. Once when my family went to a local restaurant I noticed a sign in an ominous red font:

OCCUPANCY OF THIS ESTABLISHMENT BY MORE THAN 232 PERSONS IS DANGEROUS AND UNLAWFUL. "Mommy," I asked, "what's 'occupancy'?" She told me, and I immediately began to count all the people in the restaurant. I was plagued by the thought that my family's arrival would doom us all to an awful punishment.

Now I hardly think of stop signs, but something about my childhood fascination has stayed with me. In signs I saw the natural laws of my environment. A world of fixed and simple principles appealed to me. When the rules of the stop sign and its cousins lost their infallible status, others took their place. My favorite subjects are physics and mathematics. I still feel satisfaction when I behold the universe obeying its own "signs," such as: Speed Limit—671 Million MPH, Entropy—One Way, and Quantum Leaps—Exact Change Only. These universal signs give commands that cannot be broken by careless children or reckless drivers; they are unwavering principles. I tend to see our existence governed by some simple rules written on signs posted in the very fabric of space.

When I encounter a confusing situation or a seemingly impossible task I break it down and make a mental sign with instructions for its completion. I know my method has its drawbacks. It lets me enjoy physics because of rules, but I quickly became intolerant of biology, which starts with the final products of unknown rules. I view the world in narrow pieces—a way of thinking that I know can be arbitrary and inaccurate. In the real world, everything is firmly attached to everything else. My method of rules would tell me now that I need to go beyond it to have the fullest appreciation of the world. I should probably throw away the big red sign hanging in my dorm room. Life isn't that simple.

Joseph Calzaretta received an SB and SM from MIT in Mechanical Engineering and now works in Information Services and Technology at MIT as a software developer.

CARDS

My parents bought a special deck of cards in Sweden. There were ninety-six cards in the deck. Exactly half of them had pictures of little wooden toys—boats, planes, and cars in red, yellow, green, or blue. In addition to these, there were one or two little wooden men in each picture, either outside or inside of a vehicle. Each of the cards depicted one of the forty-eight possible combinations.

The other half of the deck was abstract. Each of its forty-eight cards was divided into four boxes. One box was a square, colored in either red, yellow, green or blue. The other three boxes held illustrations in simple black and white. The first square held a simple side-view silhouette of the vehicle. A second held a picture of one or two men. A third contained a big black circle with a thin ring. It showed whether the vehicle was occupied; an unoccupied vehicle was designated by a circle with a ring beside it. An occupied vehicle had a ring around it. Once again, there was one card for each possible combination.

I would play with the cards, laying them out in big patterns, trying to create some sort of flattened four-dimensional grid on the carpet. I would take some of my own toys or balloons and put them in with the cards. I learned words like "unoccupied." I would try to make cycles of cards where each card had exactly one attribute different from its neighbor, and then try varying that rule, reducing the number of cards, or placing other restrictions on the game until I couldn't get a cycle. I split up groups based on certain attributes. I did some really bizarre number and group theory things with these cards.

None of these little games struck me as being very profound at the time. It just seemed like fun to play with the deck. I would look at the toys in the pictures and think, "Man, those wooden toys are cool and I want some." But I did have a thought that, at the time, seemed to stand out as important.

My toddler epiphany was that, even though one of the decks was much prettier than the other and much more accessible, the division between the picture cards and the abstract cards was just as arbitrary a split as dividing the deck between one-man and two-man cards. Even though I just described the deck in two halves, each with four distinguishing qualities, one could describe the deck as one set of cards with a fifth attribute. Singling out the picture cards was essentially the same as singling out the cards showing an occupied vehicle.

After that, I was at ease with the idea of symbols, with the idea of abstractions. The deck taught me to find patterns and relationships and to find relationships among possible relationships. With all the card pushing I did, the counting and the abstraction, it was pretty easy for me to learn how to read and do simple arithmetic.

When I was two years old, I heard my brother reading his old kindergarten alphabet workbooks. Later, I went into his room and began to leaf through them. Each letter had its own mimeographed workbook and mascot (like Penguin Pete or Sammy Seal), and each book started out with a simple little poem listing words that began with the special letter. I was stoked that the book on top was stapled in construction paper of my favorite color, green, so I read through that one first. I recognized Alligator Al immediately on the inside cover, and saw the "A" and "a" shapes. I figured those two shapes had to be A, and spotted them everywhere at the left side of words. I assumed that words were split up by white space. By the time my brother came back into his room to yell at me, I'd deduced that "A" was a shape that stood for the short A sound as in "Alligator" and that "a" was a shape that stood for the long A sound as in "aviation." Okay, so I was wrong, but my brother was impressed enough that he wanted to teach me how to read.

I wasn't afraid to turn sounds into letters and numbers into numerals. My cards left their legacy. I knew how to find rules and search for patterns. And I wasn't afraid to play with abstract things; the cards had made symbols into great toys.

Brian Tivol received two degrees from MIT in 1998, an SB in Mathematics with Computer Science and an AB in Film and Media Studies. He currently works as a software engineer at Google.

What We Program

BASIC MANUAL

Fred Martin (1989)

By three, I felt that I understood the Lincoln Logs world. It was finite. Things fit. The Lincoln Logs had a wonderful regularity; you could create a wall with a window by using shorter logs just as easily as you could build a solid wall. My blocks world expanded at seven, when I went to an experimental school and spent most of my time in a blocks room that had a huge collection of wooden building blocks. The basic unit of the set was a rectangular block about six inches long, but then, there were longer beams and diagonal pieces—a huge variety of compatible wood blocks. I spent days on end building complex roads, garages, and ramps with the blocks. We would test them by driving cars through the worlds we built.

The Lincoln Logs and the wooden blocks shared an aesthetic—they began with a well-defined and consistent set of first principles that enabled you to build more complex structures, so when I met the BASIC programming language I felt in familiar territory. When I first met BASIC I didn't even have a computer or easy access to one. All I had was Radio Shack's Level I BASIC manual. The manual didn't assume you knew anything about computers or had one around. It wasn't a reference manual but an interactive text that led the reader through a set of concepts, examples, and questions that taught BASIC programming.

I began to write programs almost immediately; my lack of a computer seemed rather incidental. I knew that the programs I wrote would run on one and that is what mattered. I wrote programs at home, on paper, and then waited until I had a chance to go to a Radio Shack to try them out. One Sunday, that meant a ten-mile bike ride to the nearest Radio Shack.

But most of the time, I just imagined the computer for which I was writing my programs. Because I knew this computer so well, I could debug my programs mentally and run them on my mental computer. My first large program calculated a "biorhythm chart" and displayed three sine curves on the screen. It was several hundred lines long, and the only bug it had was in the calculation of the sine waves: I didn't realize that the argument to the sine function had to be given in radians, not degrees.

Finally, when I was in the tenth grade, I got a computer of my own. I can't count the hours I spent working with it. I wasn't a very social child; having the computer took the place of other companionship. I just wasn't very interested in people at the time. What kept me at that computer was discovery and design. Now that I had this machine, I wanted to know how it worked. It did all of these complex things; I wanted to know exactly how its complexity was constructed. I wanted no mysteries to remain.

I wanted understanding for a particular type of control. I wanted the computer to do things that I wanted, but I wanted it to want to do whatever I wanted for it. In other words, I wanted to achieve a kind of merging of minds with the machine where I would not be domineering or forcing my will upon it in some aggressive way. I wanted to know how it worked well enough to be confident that it wanted to be used in whatever way pleased me.

I wrote most of my programs in assembly language, the language that kept me closest to the machine. Most of my programs built tools for the machine itself, grouping together the best of some similar products that were on the market at the time. I felt as though my programs served the machine in the most powerful ways possible. I enjoyed serving the computer, a feeling intensified by the fact that this computer was mine. I was the sole human who was giving instructions to the machine.

I came to view knowledge as alive, contributing to discovery, leading to more knowledge and to creation, the making of things that puts knowledge to work. While I explored the microcosm of computation that lived on my desk, I was engaged in scientific pursuit: information about the computer contributed to my understanding of the whole machine's functioning, just as scientific facts contribute to the formation of a model. From the day I met the BASIC manual, I began to think like a scientist.

Fred Martin is Assistant Professor of Computer Science at the University of Massachusetts, Lowell, interested in creating negotiated design environments for children, artists, and other non-engineers. He received an SB, SM, and PhD from MIT.

APPLE II

David (Duis) Story (1990)

I was in seventh grade, almost twelve, when I first saw an Apple II computer. I immediately fell in love. Within a year, I saved and borrowed enough to buy my own. In short succession, I would lose my girlfriend, spend over a hundred dollars on a monthly phone bill, have my grades drop to Cs and then have them come back to As, and grow closer to my brother without his knowing it. And I began to think of the computer as an extension of myself.

The thing that got me started was a game called *Wizardry*. It embodied the magic of computers; you could control several things at once with almost no effort. The game was an incarnation of the classic *Dungeons and Dragons* fantasy role-playing game in which a party of adventurers explores the catacombs on the outskirts of a village. They keep careful maps as they travel in order to find their way back. They fight an assortment of monsters and gain experience points and treasure.

Wizardry enabled you to create and control up to six characters at any one time, each a unique individual. For several months after I bought my Apple II, I spent almost every afternoon, evening, and weekend playing the game. I felt in total control over each character's destiny; this feeling set the tone for my relationship with the computer.

With the computer, I was in the ultimate safe relationship, distanced yet intimately involved. I was both inside the game and outside it, a player and the god who ruled, able to switch off the world at my whim. I endowed each of my characters with an individual personality: Elric, Sharra, Hawkmoon, Corum, Orodreth, and Rackhir were constantly with me. To this day I remember their marching order and their weaknesses and strengths.

My friend Eric played *Wizardry* on his father's Apple and we began a series of nightly play sessions. We would call each other on the phone and play *Wizardry* at the same time. At first we would really talk as we played, discussing our algebra homework and what we thought of different girls, but soon we talked only about what was happening on our computers, what was happening to our *Wizardry* adventuring parties. Our conversations became more and more competitive, and we used our characters' successes as yardsticks of our own worth.

> Hey! Have you been over to the troll's lair on level four? I can't figure out how to kill him. I keep having to run away.

> Man, are you a loser! I solved that *weeks* ago! Use the blue dagger, you know, the one that you get from the mage on level three? Well, just make sure you throw it at the troll. He's usually the number-four monster in case you haven't figured that out yet. That'll take care of him.

> Okay, thanks. I've almost healed all my hit points back, maybe I'll go back and try him again before I disband. Hey, I bet you haven't solved the riddle on the east end of level three yet . . .

Eric and I were both competitive people, and after our competition came to the surface in *Wizardry,* our relationship was never the same. We recognized that if we competed the same way in real life, we would stop being friends. After we solved all the problems in *Wizardry* we avoided all competitive situations.

The phone bills were also a problem. My mother informed me that I would have to pay in full the toll charges on calls I had made to my friend. I had spent $137. After that, Eric and I called each other only after 11 p.m. when the rates went down. We came up with a scheme of brief rings so that we could call each other late at night without waking up our parents. We even arranged for a friend of mine who lived halfway between us to have his phone forwarded to Eric's house so that I could avoid the toll charges. I bought a neck cradle for my phone so that I could type with both hands and still talk on the phone, and soon after, I became the first teenager in town with a speakerphone. I lived on the phone.

And I lived for my computer. It was always in my thoughts. Whenever I was away from it, I wondered what I could be doing with it. If I was in the room with it, I had to be using it. If I was in the same house as it, it had to be on, doing something. I had a hard time staying out of my room. I began to wolf down my food at meals, finishing dinner almost as soon as my mother put the food on the table. I lost ten pounds without noticing it. I stayed up past midnight so often that I regularly overslept and began missing the school bus. I hitchhiked the fifteen miles to school at least once a week.

One morning at school I was hurrying to finish my algebra homework during recess when I realized that my girlfriend Patti was standing before me. She was my first real girlfriend. We had been good friends for a year and had been going steady for almost nine months. At this point, I had owned my computer for about three weeks. She was crying, but when I stood up to calm her down, she stepped back and said: "I don't know why you haven't been paying any attention to me since Valentine's Day, but I can't take any more. We're through."

I am struck now that what could have been a traumatic event (and certainly was one for Patti) barely made an impression on me. I hardly gave the breakup a second thought.

One day my sister asked me what I had named my computer. I had not done so, and she was shocked. But I remained firm on the point. Later I realized that if I named my computer, I would be endowing it with a personality, and it would start having problems. I got this idea from my observation that when people named their cars, they attributed personalities to them and that their cars then became problematic. Therefore, my computer would have more problems if I allowed it a name. I felt that if I let down my guard for even a moment, I would let a name slip, my computer would develop human problems, and it would take months, possibly forever, to repair the damage.

As I worked on my computer, my role model was my brother Jim. He was a genius with motorcycles and cars. He knew how anything and everything mechanical worked. I respected, even revered, him for his knowledge. One thing was certain: Jim never named any of his machines. And Jim never swore or cursed at them but calmly and rationally analyzed any problem that came up and then set about solving it. I decided that the people who referred to their cars as people did not understand the inner workings of their cars. I did not want to be like them. I wanted to be like Jim. I adopted my brother's disdain for the mechanically disinclined. When I was with my computer, I could be like my brother. I could understand something completely. I could be an authority on questions about it. I could remain calmly in control when faced with problems that would make other people lose control.

Control: I was the son of an alcoholic father who had left my mother when I was ten. I was the third child, eight years younger than my siblings, who were fraternal twins. I didn't feel much in control, but the computer gave me a place to have that feeling. The consistent response of the computer provided me with the assurance that my world was secure and that any work I did would remain forever. Or perhaps it was a consistent companion, one who could respond to my wishes and not desert me.

Soon after, I purchased a modem, a device that allowed my computer to transmit data over telephone lines. I began to meet people via modem by sending text messages to computer bulletin boards. This was a new experience, friends that I would never meet or even speak to. I also used my modem to transmit and receive games. Now I had a new task for my computer. Now it was easier for my computer to be in use at all times. I tried to make sure that my computer was always receiving or transmitting something—a game, a word-processing program, whatever. What was important to me was the feeling that my computer was busy—because only when my computer was busy could I pay attention to something else.

So, for the first time in the six months that I had had the computer, I could be in the same room as the machine without having to have my hands on its keyboard. The transfers of programs would take up to an hour each, so I was now able to concentrate on something other than interacting with the computer. My grades jumped from Cs back to straight As.

I had become obsessed with occupying my computer's time. I felt that I was wasting something precious if I didn't have my computer working all the time. Reflecting back on it now, I think that I was satisfying my own desire to be productive and lazy at the same time. I could be lazy if my computer was being productive. It was my mechanical prosthesis.

Once I began to have some time to myself while my computer was busy on its own, I started to realize that I had been missing out on a lot. I took a job that kept me away from the computer from right after school until dinnertime. My obsession had come to an end, but the attraction has lasted until this day. I began to see my activity as that of both myself and the computer. I was productive if it was productive. Despite my insistence that my computer not have a personality, it became integrated into my own.

David (Duis) Story graduated from MIT in 1990, majoring in Computer Science, and then worked for Silicon Graphics Incorporated and Intuit. He is now Vice President of Engineering for Digital Imaging and Web Products at Adobe Software, working on Photoshop, Lightroom, and Dreamweaver.

ATARI 2600

Ji Yoo (1994)

One of my aunts bought me an Atari 2600 Home Video Game Center for my ninth birthday. My parents, being parents, would have never bought me something that cost $150, with cartridges at $30 apiece. I hurried to the television set we had in our playroom, and quickly reading the instructions, set up the computer. For the first few days, my brother and I furiously cycled through the cartridges, playing game after game, not doing or caring about much else. It was like eating a few gallons of ice cream and not getting sick afterward. The more we played, the more we wanted to play. We neglected our homework and friends.

My parents became concerned about the nonstop activity, so they set up some rules: two hours on weekdays; three hours a day on the weekends. My brother and I divided the time between us and took turns. The first game that completely engaged me was *Missile Command.* As player, you had to defend six cities from missile planes and smart bombs. Your defense weapons were three silos that each had ten laser beams.

As I conquered each layer of the game, things got faster. The missiles and bombs fell at a near-frantic pace; my hand was getting cramped from moving the joystick so quickly. Gradually, I understood that the game would continue to get harder and harder until it would surpass what was humanly possible. Finally, no one would be able to complete a level no matter how hard one tried. My solace was that I could take pride in being the best *Missile Command* player on my block.

Before the games, the kids in my neighborhood would have good-natured contests for free-throw shooting, bowling, and spitting-for-distance. The games added a new dimension to our competition. My friends and I stopped playing with LEGOs and began to take turns playing a different game every day, going for the highest score.

With traditional outdoor contests that took place on the playground, we would set some sort of limit ("best out of fifty-five throws," for example) and stick to it, and then go on to do something else. With video games, we fought instead—and all the time—about whether to set any limit on the number of games in a contest. Some kids thought that it would be enough for each kid to take one turn at each game. Others felt that three games per person would better represent the skills of each player. We could never agree and settled on playing as many games as we could until my parents pulled the plug. When we played, we were each in our own separate world, oblivious to anything else. The kids who were not playing would fidget and mumble, trying to distract the player so that their turn to play might come faster.

One of our favorite games for competition was *Kaboom!,* a game whose object is to use baskets to catch bombs that a "mad bomber" drops from the top of a wall. At the early levels of play, the game was simple, but as you advanced, the game challenged one's reflexes to the limit. The mad bomber frantically skittered across the top of the screen, dropping bombs at a maddening pace. At the very high levels of play, I lost consciousness of myself. I wasn't thinking of what I was doing, I just did it. Like a piano player's fingers that can "run away with the music" and seem to have a mind of their own, with practice I was able to move the buckets as fast as the mad bomber was dropping the bombs. My fingers moved without my conscious direction. There seemed to be a direct connection between my eyes and my fingers, my brain a passive observer.

On the rare occasions that I played alone, my favorite game was *Pac-Man.* I read up on the game and was able to develop a pattern for my play that would enable me, as Pac-Man, to gobble the dots in a predetermined sequence while the four monsters that were chasing me wandered around the maze, frantically trying to catch me. No matter how difficult the game got, no matter how fast the monsters chased me, my pattern always worked. Four monsters could chase me, but they never would catch me. When I realized that this pattern gave me the power to play this game indefinitely, I felt as though I had transcended the game and become one with the computer.

I was, in short, addicted. And I was able to understand my parent's worst habit in a new way—the fact that they smoked. I put myself in their place and equated smoking and playing video games. I imagined smoking, feeling a brief sense of relaxation, and then, an hour later, the urge to smoke again. I imagined knowing that smoking was bad for my health and yet being powerless to stop. By thinking about the games and smoking, by seeing them as the same thing, I decided that I needed to break my addiction to the games. I managed to curb my playing time, but I was never able to break my dependency.

To try to distract myself from games on the computer, I tried, with my parents' help, to see how the game machine actually worked. I read electronics books that were way over my head, but after a while I was able to get a basic idea of how the game cartridges interfaced with the game console. By this time, more sophisticated computers than the 2600 had come out, with faster CPUs and disk drives. These computers could store information and have it be altered in any way that the programmer saw fit, unlike the cartridges that could never be changed. I envied the owners of these new machines. When I got bored of any game, there was nothing I could do except ask my parents to buy a new one.

I became taken with the idea of opening my game console and cartridges, even though I knew that this risked permanent damage to the components. One day, my curiosity overwhelmed me. In secret, I took my dad's screwdriver and opened the console. I was surprised to find a complex and confusing array of wires, circuits, and other whatsits going every which way. I realized it would take me a long time to decipher what each of the parts did. With the help of a circuit book, I would open up the console and try to figure out what was going on. I realized that the world of the game, that I had seen as so simple and clean, a few blips on the screen, was underneath, a world of great complexities.

My involvement with the Atari 2600, game playing, and circuit exploration came to an end when one day I tried to play *Missile Command* and the console did not work. I realized that during one of my secret electronics lessons, I must have shorted a circuit on the main board. I did not know which one it was and could not repair it myself. I did not tell my parents the whole truth; I simply informed them that the game was broken. They reluctantly agreed to have it fixed, but while it waited in the car to be taken to the repair shop it was stolen at my school during a PTA meeting.

Ji Yoo graduated from MIT in 1994, from medical school at the University of California, San Diego, in 1998, and finished his psychiatry residency in 2004. He is currently working as a psychiatrist in the US Navy, stationed in Yokosuka, Japan.

TRS-80

Chris Dodge (1996)

By the time I was nine, I had read a computer book that included computer games written in the BASIC programming language. Playing these games was boring; what interested me was the relationship between the game and the representation of the game in a program. I tried to form connections between results in the game-playing space and the program code. There was clearly a cause and effect that was being outlined in this abstract language, although I could not understand it. I devoured my computer book, even though I didn't have a computer or access to one. I was fascinated by the possibilities of building a program.

After a hard-working autumn raking leaves, I earned $200, and for my Christmas present, my parents and I split the cost of a TRS-80 computer. I was overjoyed when the computer was out of its box and set up on my bedroom floor. It was a scrawny looking thing; it didn't look much like a piece of modern technology. It had 4K of RAM, 160 × 160-pixel monochrome graphics, and a tape recorder for its storage device that ran at 250 baud. None of this was of concern to me. I could finally apply all of the intuitions about programming that I had so painstakingly accumulated. Without much hesitation, I chose my first program and typed it in. It was a novelty: it showed a cannon perched atop a castle firing a cannon ball, making a small graphical arch across the screen. I didn't care if it was silly. For me, the purpose of writing computer programs was not to accomplish a task or play at a game, but to explore problem solving and cause and effect.

On the evening I completed my cannon program, my grandfather, a stubborn and erudite old man, decided to compete with the computer at calculating simple arithmetic problems. Or rather, he competed with the computational potential of the machine. I was the medium between my grandfather's world and the spirits that lay within that awkward-looking gray box. I acted as programmer; I coached out the computation. Without me, the TRS-80 would have no consequence; it would be left to stare out frozen at the world through an inert command-line prompt.

The TRS-80 was incomplete, and so it was perfect. It needed an operator to act as God, giving Adam the breath of life, just as G.I. Joe needs a child's fantasy for it to have significance. As I worked to make the computer compete with my grandfather, there was an instant in which I understood what it means to be a programmer. It was a powerful, seductive moment. As programmers, we are the circus ringmasters of a parade of symbols that represent our understanding of the world. We map one reality into another, hoping that the projection will be, if not accurate, at least a worthy approximation. In that short moment, I was transformed from an awkward ten-year-old to a keeper of an arcane knowledge that bubbled beneath the keyboard. In my mind, I became a monk, a scribe, encoding the understanding of our times into the machine for results to be returned to humankind. The machine became as a mirror of my beliefs of the world. It was a harsh mirror, like the glare of a fluorescent light in a Texaco gas station bathroom. And through the eyes of my grandfather, I had changed from a grandchild into a communications channel to the threatening unknown.

Eventually I set up the TRS-80 on an old, rickety card table inside my walk-in closet. This is where

I spent my young years, in the closet. I would not have appreciated the aptness of this metaphor, because in those years, computers were not cool. A serious computer hobby was well kept hidden. These days, many of the richest and most powerful people on the planet are computer geeks, but in the early 1980s, computer aficionados were not held in high regard. I was encouraged by my parents, but without the sense that this was the start of a career. There was little opportunity for validation from my parents. When I showed them what I was doing, they were uncomprehending.

However, it was in that closet that I developed my programming skills into a good intuitive understanding of problem solving. The primitivity of the technology forced me into a constant struggle to accomplish very simple tasks. BASIC was the only programming language available. There were severe limitations in speed of processing and amounts of memory. Most difficult was the tape recorder. It was slow and unreliable. And it came with its own catch-22. It was not possible to verify that a program was successfully saved unless one tried to play it back, but when one tried to load a program, the previous program was erased. So if the version that had been saved on to the tape was bad, one no longer had the original. This limitation ate up many days of work. But none of this was in the least bit discouraging to me because it was the process of programming, not the utility of the software that so engaged me. After losing a day's work, I shrugged my shoulders and started fresh on the next day.

I tinkered mightily to find solutions that worked around the machine's limitations. One idiosyncrasy of the TRS-80 made it possible to use the keyboard while a program was running. This left a visible mark on the screen that sometimes disturbed the displayed graphics. I developed a mechanism to allow for real-time keyboard input to a *Space Invaders*-style computer game I wrote. Although the TRS-80 Model I, Level I, did not allow for real-time keyboard input, such as the arrow keys to be explicitly recognized by the program, I was able to come up with a work-around. If the user hit the left arrow, the cursor would back up one space; if the user hit the right arrow, the cursor would go forward one space. If the user hit the enter key, the entire line would be erased by the carriage return. So I placed graphic dots to the left and right of the cursor and one at the end of a line. The *Space Invaders* game would "look" at these dots to see if they had been erased by the cursor; that would indicate that the user had pressed an action key. (If the left dot was erased, the user must have hit the left arrow, and the user's spaceship would move to the left; similarly, if the right dot was erased, the user must have hit the right arrow, and the spaceship would move to the right. If both the right and left dot had been erased, the user must have hit the enter key, and a bullet would be fired.) So, the impoverished TRS-80 brought me into the spirit of problem solving: one begins by discovering the boundaries inherent in a system and then devises a solution that circumvents these limitations.

Gradually the novelty of such creative kludges wore thin, and I moved on to programming in assembly language. The process of problem solving was the same, but I felt an emotional shift. With BASIC I felt a playful glee in tinkering. When I met assembly language, I felt awe and—without any exaggeration—fear. It was as though I had witnessed the process of creation alongside the Creator. I cautiously entered in lines of code, conscious that with any false step I could crash the machine and lose everything. When I programmed in BASIC, the computer seemed an indulgent parent. The programmer works within a protected environment, distanced from the harsh realities of the computer itself. Working in assembly language felt like working with the real thing. I was in the underworld of bits, registers, AND/OR gates, interrupts, and signals. I was in the fully deconstructed reality, the end-stop of real-world representation.

When programming in assembly language, the world exists as a set of concentric surfaces, much like an onion, each layer working on top of and dependent on the lower levels. Assembly language brought me to ideas about general systems such as scale, representation, problem abstraction into physical and abstract layers, and the idea of scope. I began to think that software construction with such ideas is a general model of how systems are organized—even systems outside the computer, such as government, economics, and physics. I made analogies between evolution and levels of computation. As I saw it, as evolution progresses, people move out from the central core of what a machine is and use a shared history of interaction to build more complex and higher-level representations of the world.

I used the computer to model the world. It became my conceptual seed for thinking about complex representation. I tested my worldviews against what I could model on the computer.

When I look back at my TRS-80, I think that my moment in time was unique. I was given a chance to tinker with a machine in a way not easily possible today because that moment of "hobbyist" activity requires a system that is "open"—meaning that it is possible to "open the hood" of the computer when necessary to fiddle with some underlying component of software. This tinkering relationship with the computer can happen only if the underlying software is relatively straightforward. The tinkerer needs to have a good solid overall understanding of all the lower-level subsystems of the machine. As computer software engineering becomes ever more sophisticated to meet the demand for user-friendly applications, what stands behind the opaque user-interface becomes too formidable for any one person to understand.

I feel a loss as I become dependent on other people's programming tools, other people's layers. I no longer can tweak and tinker as I did as a child. These days, my programming style is superior, and I can accomplish many utilitarian tasks in an efficient manner. But the TRS-80 required creativity even to perform menial tasks. I have recently come to the conclusion that I did my best computer work between the ages of ten and fourteen.

Chris Dodge, an experimental video artist, graduated from MIT with an SM in Media Arts and Sciences in 1997. Since his graduation, he has worked in several Boston-area startups in the field of online digital media.

BASIC

Nelson Minar (1996)

I first read the Applesoft BASIC manual when I was eight years old. Despite the fact that I had no access to a computer, I read it over and over. I imagined what I could do in BASIC. I learned that people could control the magic inside the box by writing programs, that people could create their own wide worlds within the computer.

I began to scavenge time on computers, hanging out in computer stores, staying after school in the special computer room, going over to friends' houses. I had a favorite program on the Apple II, *Lemonade Stand.* It was a game in which the user controlled the business of selling lemonade, making choices about investments and risk. Since I had read the BASIC manual, I knew that it was possible to control this virtual world. After playing *Lemonade Stand* as a game, I would stay at the computer and list it out as a program, reading it and trying to understand it. I reached inside the program and cheated: my friends and I had built a *Lemonade Stand,* and I gave it infinite funds. Of course, this took the fun out of the game, but it impressed my friends. More important, I had the experience of taking apart someone else's BASIC program and realizing that I could figure it out.

I played many computer games; my favorites were the role-playing games such as *Ultima* and *Wizardry,* games that offered virtual worlds to explore. I spent hours inside these games, playing my character through the story, appreciating the neat things the game designers had created for me. But I was never entirely content to play within the games. I always wanted to get underneath and understand what was really going on. I figured out how to take pieces of the games apart, read through their data files to find out everything that could happen, and modify things to make myself more powerful. I tended to do this only after I had gotten bored with playing the game itself. Once I understood the surface I was supposed to play on, I would make a game new by taking it apart to understand how it worked.

So at the same time that I was navigating artificial worlds, I was also learning how to write my own programs. When I was about twelve I had gotten tired of programming in BASIC and set about learning machine language. I did this because I had been told that machine language was more powerful, that all of the coolest games were written in it. I quickly understood that machine language was the language of the computer; it was a machine language program that turned BASIC programs into machine language instructions. I immersed myself in *What's Where in the Apple I*, which offered a disassembly of the Applesoft interpreter. This book explained how BASIC worked as a collection of machine language subroutines. I felt that I had finally uncovered the true reality of the computer's magic.

I wrote many programs in machine language. To me it seemed that I was working my own magic to create worlds. One day, however, I was rather idly writing down bit patterns for all the machine language opcodes (strange, what kids do for fun), and I noticed that the bit patterns themselves had a structure, an organization. I had uncovered the microcode inside the microprocessor, the program run by the machine language instructions themselves. I remember being pleased that there was yet another level of organization in my computer, happy that things made so much sense. But I also realized that I did not really care, that I needed to understand only the level of machine language to do everything I wanted to. Why didn't I care about microcode? Because I couldn't change it: I could not write my own microcode. I was not able to use it as a tool to build software. My goal was to create things in the computer. For that, machine language was as deep as I needed to go.

From all this tinkering I learned a concrete, direct understanding of functional abstraction. At the highest level of abstraction there were the computer games: worlds you played in, following the rules of the game designer. But these games were themselves written in a computer language, and people program their computer languages to tell the computer what to do. These computer languages were in turn written in other computer languages, down to the microcode. Each layer of abstraction defined its own artificial world. Inside a computer we build worlds on top of worlds.

My direct experience with functional abstraction organized my understanding of the natural world. I see the relationship between branches of science—for example, between physics, chemistry, and biology—in terms of layers of abstraction. Chemistry requires the understanding of a lot of details about molecules: how atoms bind together to make objects with certain properties and shapes, how those molecules interact. But at the level of cell biology you can choose to ignore a lot of this chemical complexity. You can work on the biological surface.

These days, as a biologist, the systems I work with are seldom as straightforward as the layers of a computer system. Details of chemistry intrude on the tidy world of biology, but the basic capacity to organize complexity, to put it in boxes, is essential to our ability to understand the world. The idea of functional abstraction is at the center of my research, as I study emergent phenomena in complex systems. I work on software tools to help me understand how pieces of a natural system might come together to form a whole unit, how complexity might emerge from simplicity. As I work to design a system at one level so that it possesses coherent functionality at a higher level of abstraction, my early experiences inside computers continue to shape my understanding of the natural and artificial.

Nelson Minar studied at the MIT Media Lab, founded a distributed computing company, joined a large Internet company, and now develops tools for programmers.

ATARI 800

Steve Niemczyk (1996)

In 1981, the Atari 800 became the first computer to enter the Niemczyk household. I remember the day my father brought it home. I was seven. The shiny, brown-shelled object was placed on a humidifier and my father connected it to our nineteen-inch Panasonic TV. Like all other technological objects, it would be under my father's watch and care. I stared in fascination, as my father inserted the first cartridge, *Star Raiders.* Our family already had the Atari 2600 game system. I felt familiar with what cartridges could do, but my dad said that this was a *real* computer, and so I thought its games would be better than I had known. My dad fumbled for the switch and out of the snowy chaos emerged the large blinking letters: STAR RAIDERS. I began to dream of what the game would be like. It would be spectacular. It would let you fly from planet to planet, conversing with aliens. It would let you be Han Solo or Luke Skywalker. It would let you rescue the princess and negotiate with Imperial Officers. It would let you learn the ways of the Force and decide for yourself whether to sway to the Dark Side.

In fact, the game did none of these things. When my father handed me the controls I quickly became bored and lost the game. I asked my father what other cartridges came with the computer, and there was only one other. It had the peculiar name BASIC. We inserted the cartridge. A blue screen emerged. My father typed:

```
READY

10 PRINT "HELLO"

20 GOTO 10

RUN
```

To my amazement, the word HELLO scrolled endlessly across the screen. This game seemed silly, but my father explained that this was not a game but a programming language. And he said, "You can make it do anything you want." This fascinated me. My mother had a book that explained how this BASIC thing worked, and I set out to read.

I was amazed by the power of this simple device. I could make it say whatever I wanted. I could make it do whatever I wanted. That very first day I began to make my own games, my own creations. Programming felt like a liberation from the confined world of my elementary school in suburban Long Island.

But there were obstacles. My father feared that as a young child I would break the computer, so I had to design programs on paper before I put them on the computer. Only when I was sure a program would work, or rather, when he was sure it would work, could I use the computer. But from the small amount of time I had spent at the machine I already knew two things: first, programs, no matter how carefully created, will have flaws; second, I learned best through interacting with the machine. To assure that a complex program would work, I had to get inside the machine and assure that every little piece did its job. Only then could I create a unified project. Both were at odds with my father's rule, so I had to work around it. When I worked on a project, I created a fake program for my father, one that looked correct, but that contained only a superficial part of the project I intended to create. I had already realized that my father did not know whether the program I handed him would work. I had discovered the fallibility of my parents. In one year I had developed a skill that exceeded theirs. I was only eight.

When I was ten, my father moved on to other, more advanced computers, and I inherited the Atari 800. Now I could use it however I wished. The do-it-in-advance rule was no more. I created my own video games and taught myself mathematics that I would not see again until high school or college. Abstract concepts such as variables, functions, and parameters seemed concrete in the context of computer programs. I learned to solve math problems by creating algorithms, and I learned to check whether my answers made sense. I could even think like the computer; I could use procedures and debugging to approach problems in the world beyond the computer.

What I loved was that, with the Atari, I could know what every single byte address in the machine did. I actually owned a 200-page book that went through each one: *Mapping the Atari.* This book listed every place to "peek" or "poke" within the computer and told me its function. When newer and larger computers came into the house, I wondered how I would memorize what all their addresses did, the way I had mastered the Atari. Soon enough I realized that this level of knowledge would not be necessary to operate the computers I would use in the future. But the love I had for the Atari 800 will never be exceeded by another computer. I am a computer scientist at MIT, but the feeling of control over a machine—that feeling of total knowledge—is something I cannot find with any computer of today.

Steve Niemczyk completed an interdisciplinary doctorate at MIT in 2002, focusing on new collaborative instructional environments on the Web and, after graduation, was Principal Investigator for an NIAID grant to investigate hospital outbreaks of infection. Currently, he works at OPNET on algorithms that identify causes of network bottlenecks.

APPLE II

Rachel Elkin Lebwohl (1998)

I love computers. I use them to solve problems; my favorite jokes are about people who program them; I think and occasionally speak in programming languages; sometimes I even imagine that I *am* a computer. But when I recently caught sight of my first computer, my family's old Apple II, lying on the floor of the garage next to a defunct toaster oven, my heart went cold. This was no old flame, no first love. Sure, that old Apple II brought back childhood memories of typing up reports with the Bank Street Writer, playing the occasional game of *Centipede* or *Frogger,* and generally enjoying the good old days when floppy disks were floppy. It seemed almost impossible to believe that I hadn't once loved the thing.

But I hadn't. It was always my older brother Carl's domain. After school, Carl pushed the Logo computer language to its limits, just for fun. He and my dad would sit for hours together at that Apple II. My dad would encourage me to join them, and they never actually excluded me, but the big square box with beige paneling was always "Carl's thing."

I do have one proud memory of programming at the Apple II. When Carl discovered BASIC, he wrote this program: "10 PRINT Rachel is stupid. 20 GOTO 10". I was indignant and wanted to fight back, so my dad, who always insisted on getting at the heart of the matter, used the moment to teach me the concept of an infinite loop. So I wrote: "10 PRINT Rachel is awesome! 20 GOTO 10".

I never went on from there, because Carl stopped writing taunting programs. It was years before I felt that special sense of a programmer's accomplishment again.

Because computers were "Carl's thing," that Apple II almost taught me that computers were not for me. Luckily, my father and I spent many dinners working out interesting math and physics problems on paper napkins. My physics teacher encouraged me to work with a mentor at General Telephone and Electronics, GTE, who taught me about this quirky and beautiful thing called the Internet. This led me to a summer programming job, which in turn led me to major in Computer Science. My college classes led me back to infinite loops, to more useful algorithms, to experience writing and debugging complete programs, and finally to that special pride: seeing a program I'd written do what it was supposed to do. When I felt it, I recognized that I had first experienced it at my brother's lovely, now lonely and nearly forgotten, Apple II.

Rachel Elkin Lebwohl studied Media Arts and Sciences at MIT and is a software developer and project manager at NYU Medical Center.

MODEM

Anthony Townsend (1999)

I was about eleven when I heard about modems and immediately was among the converted. However, it was going to take more than a few good days at the lemonade stand to save the $300 to $500 necessary to buy one of those beauties. Complicating the picture was that my computer, a Texas Instruments 99/4A was a discontinued model. There was no guarantee that a modem would become available to work with it.

A few years later, my family moved from a New Jersey suburb of New York to a sandy shore a few hours south. After twenty years of weekly business travel across the country, and missing the troubled adolescence of my two brothers, my father had decided it was time to settle down. But living at the beach is not the same as visiting the beach. The end of the summer in a beach town is like the end of a party. Life on Cape May was very isolating.

On my thirteenth birthday, my parents surprised me with a new computer, the Tandy Color Computer 3, sold through Radio Shack stores. Radio Shack was one of the few stores I could reach by bike, and living at the shore meant that during the summer, even a thirteen-year-old could forge working papers and clear $1,000 dollars for the season. The manual that came with the color computer listed one peripheral device I could not live without: the DC ModemPak. When summer came,

I got a job washing dishes and earned enough money for the modem, but it took months before the local store could find one and have it shipped.

In the meantime, I had compiled a list of four or five bulletin board systems that were a local call from our home. There were not many; in 1986 only a handful of people on the Jersey shore had the skills, time, and dedication it took to operate a bulletin board system (BBS). BBSes were a kind of proto-Internet that sprang up around the United States during the early 1980s, an ad hoc network of personal computers that would take incoming calls from users like me and call each other in the early morning hours to exchange emails and public messages, handing them off across the country much like the Pony Express. A BBS with more than one incoming line, so you could page and chat with other users, was considered highly interactive.

The ModemPak looked as though it was supposed to plug into the Color computer's single cartridge slot. But the floppy disk drive interface was already plugged in there. I had anticipated this from the pictures, but I had hoped so much that my computer could reach out to the world without losing any of its capabilities. No such luck. I would not yet be able to download and save programs and files to my computer as I had read about.

To call a BBS with the ModemPak I had to pick up the telephone handset, dial the computer I was trying to reach, and wait for its modem to answer. Half the time, the line would be busy, so I would have to hang up, wait, and try again. When the modem finally answered, I would punch the ModemPak's "connect" button and hang up the phone while the computers screeched at each other to figure out who was saying what. After the dialing and typing and screeching, what appeared on my screen next was nothing less than magical:

WELCOME TO REALITY ALTERATIONS BBS. PRESS [ENTER]

This scrolled across the screen at 300 bps, which takes a full second to write each 32-character line my screen could display. It was almost as if someone were out there typing this information personally onto my screen. No experience with computers since has been as thrilling as what I felt with those first BBSes. When text comes in at 300 bps, it is slow enough that you can see the letters appear one by one. We are long past that now, but it gave me the feeling that I was making a connection, communicating, however slowly.

Before long, I had accounts on a half-dozen systems that were within a local call. Because each had only one incoming line, my time on the system was usually limited to a half hour a day. So, an evening's "surfing" meant hopping from one system to another until I had exhausted my time quota on all of them. I would stay up and online far past my bedtime, having memorized the commands and keystrokes to call, login, and navigate a BBS, guided only by the glow of the monitor. The online world meant information that wasn't available in the dull, off-season resort town that I lived in. The old nerds had BBSes with names like "Dave's Place" or "Downtown." The teenagers, who in reality lived in dull beach towns such as mine, hung out in BBSes called "Reality Alterations," "The Hotel Royale," "Inferno," or "Tao."

Via the modem, a new kind of information was coming into my home, into my very room as the single red light-emitting diode on the ModemPak winked on and off to announce the receipt of each new bit. As I absorbed all the local systems had to offer, I cast my net wider in search of obscure bits of technical information about my computer or about some musician my brother had told me about. Each new BBS's login screen displayed the numbers of new systems that were within a local call of it. I felt my way across the nascent datascape of New Jersey and onward, toward the cities, feeling the rush as each system held more information, more and more sophisticated people.

That humble ModemPak changed the way I thought about the world. With the unquestioning acceptance of a child, I came to understand the emerging world of computers that could talk to each other, the proto-network of personal computers linked together on networks such as FidoNet that would eventually join the Internet in a great mind meld of hobbyist and academe. I came to accept it as completely normal that I could not be someplace physically but could be there electronically. I drive past the Oradell exit on the Garden State Parkway when I go home to see my parents, and I remember a system, 'The Rainbow Connection," that I used to call there. But I have never set foot in Oradell. I don't even know how big it is or precisely where it is located.

Anthony Townsend received a PhD in Urban Studies and Regional Planning from MIT in 2003 and a Fulbright Fellowship in 2004 to study social impacts of broadband communication in South Korea. He is one of the founders of NYCwireless and is a Research Director at Institute for the Future.

TURBOGRAFX 16

Antoinne Machal-Cajigas (2007)

For twenty-six years, my father worked as an electrical engineer for the US Army. Electronics were his passion. They filled our home. There were always things around the house that he had built. One of my first memories was of one such project, the first piece of electronics I ever fully grasped. This was a circuit that, when powered by a battery, asynchronously lit two light-emitting diodes (LEDs for short). The circuit was stamped out on white printed circuit board, PCB for short.

When I held the white PCB and battery in my hand, I could make the LEDs flash for as long as I wanted. What I saw at five were different components: small cylindrical things with stripes, metal cylindrical things without stripes, and lines that connected the components to each another. Later, I would learn that these components were resistors, capacitors, and wires. I understood that batteries were required to make electronics operate, but I didn't understand exactly what they were doing. Although there were many things left to be understood about the circuit, I knew that I controlled it, that I could make it work by connecting the battery to the circuit. If I didn't place the battery where it needed to go, the LEDs would not flash. I played with this circuit day and night. I brought it in for kindergarten show-and-tell. I was proud to explain how it worked.

I had never been as consumed with an object as I was with the white PCB. I had model cars, race cars that you could run on electric tracks, books, balls, and even a bike. But it was the PCB that was the beginning of my obsession with electronics.

I began to examine other electronic devices around my house. I searched for circuits by looking for PCBs. I could see that VCRs had them; they were electronic devices. My father showed me the PCBs inside our radio— he took off its back cover. It was an electronic device as well. Each new discovery was an important as the last. Each discovery added a new entry to the list of Things with Electronic Parts. Despite my desire to learn about electronics, I never took any equipment apart to see its inner workings. I gathered information by observing what would happen to the output as I varied the input.

I tried to use this way of thinking with the TurboGrafx 16, the marvel of the gaming industry when it was introduced to the US market in September of 1989. My parents bought one for me as a Christmas present when I was about to turn six. Its game cartridges were no larger than a credit card and as thick as three stacked together.

My father helped me set up my TurboGrafx 16 to its input and output devices. First, we hooked the console to the power supply. Next, we connected the console to the television, its output. Finally, we connected the TurboGrafx 16 to its input, the controller. My father explained that the machine needed information to perform its tasks: "You cannot know that I want you to get me a glass of water right now unless I tell you so."

For the first time, I wanted to know more than what I could learn simply by observing the output as I varied the input. I set out to understand my TurboGrafx 16. I tried to draw analogies to the white PCB I knew so well. In the PCB, many components connected to each other between the battery input and LED output. Was the same true for my console? What components did it have inside? How many components were there?

What was I doing to these components as I pressed the buttons?

I looked for clues. In game called *Bonk's Adventure,* pressing one button produced many different actions. When the character Bonk is in water, pressing the II button would make him swim. When Bonk was walking, pressing the II button would make him jump. When Bonk was in the air, pressing the II button would make him spin. To me at six, this was worrisome. Although I didn't know how the console worked, I did know that the machine could take in information as input and produce information on the TV as output. But how could one input lead to so many outputs? I was troubled. It could not be in the pressing of the II button. If the II button was like the battery in my white PCB, pressing it should always bring about the same result. My first effort to understand circuits by treating them as black boxes, by looking at inputs and outputs, left me with no other possible explanation of the workings of the number II button than to think that the machine knew what I wanted to do. *Whether I liked it or not, I had to accept that my TurboGrafx 16 could read my mind. And if this was the case for my TurboGrafx 16, then other electronic devices could do the same.*

This mind-reading hypothesis felt like a true revelation. I thought I had come to a breakthrough in electronics, but curiously, one I did not share with my father. After a while, I grew dissatisfied. I decided that input/output analysis would not get me where I needed to go, that mind reading was an improbable hypothesis. It seemed to me that more specific questions might lead to more scientific answers. I refined my queries. How did the fifty flat rectangles at the end of each plastic card store the information for each computer game that I played? How did the reality represented in a game, including geography and characters, get stored in a space as small as the end of a plastic card? I looked up to my father. People like him knew the answers. Knowing that my father made the white PCB meant that someone else had to have made the console. Some people understood machines well enough to make them do what they wanted. I could treat certain levels of machine processing as a black box, but there was a limit. What I needed to learn was what those limits were. I needed to program. What I wanted was to understand electronics well enough to make my own.

Antoinne Machal-Cajigas received an SB in Electrical Engineering and Computer Science at MIT in 2007.

PART II

Mentors and Their Objects

What Made a Scientist?

What We See

MICROSCOPE

Susan Hockfield

How do you understand how something works? From as early as I can remember, I wanted to see inside things, to understand how they worked by understanding their internal structure. The organization and scale of things has always held a fascination for me. I picked up magnifying glasses all over the place so that I could see things at high resolution, close up. Throughout my life, I've magnified things and taken them apart. Even in elementary school, I remember fiddling with a door latch at my friend's house, impervious to her pleas, "Please don't do that; you always take everything apart." It was true. I took apart anything I could to understand how things worked. Later, this interest led me to graduate work in anatomy, with a specialization in neuroanatomy, the study of the brain through its structure. An anatomical perspective lets me use structure to understand function.

I recall a small experiment when I was about eight or nine, when I dissected a watch—one of the old-fashioned, pre-computer varieties. At some point when you're dismantling a watch you get to the main spring, and if you don't know what you're doing (which I didn't), the watch explodes. Even in an unwound state, the main spring of a watch is pretty wound up. And when you release it, the parts go flying everywhere and there is no hope of getting that watch back together, ever. Undeterred, I continued with household objects—an iron, a toaster, a fan, lamps, a vacuum cleaner.

At the same time, I found miniatures of all sorts fascinating. Trains, dollhouses, models that represent the world in small scale. I never had the dollhouse of my dreams, but then, I never saw the one I imagined, with electricity, running water, a heating system, plumbing, all of the mechanics. I wanted a fully functioning miniature, so that I could understand how a house works. But I imagined a dollhouse and a fabulous train set, with the trains traveling through mountains, loading and unloading materiel, a train set with many moving parts and equipment to understand, to get inside of.

Finally, finally, when I was ten, my parents gave me a microscope. I used it to take all kinds of things apart. I examined everything I could get my hands on, to see them enlarged, so that I could dissect them in my mind and understand how they worked.

My microscope did not disappoint me. What I found most exciting was the feeling of moving across levels of structure and scale to understand function. As a freshman in high school I took my first biology course and studied transparencies of layered acetate depicting the structures of various animals and tissues. I loved the perspective afforded by these pages; through them I moved from skin, to muscle, to organs, and to bone; some even took you inside tissues, revealing the internal structure of bone with its lacey trabeculae, or the incredible organization of the kidney that filters waste from the blood stream. I made a few biology projects with layered acetate and realized that my real interest was in living things. The medium of acetate layers offered the same kind of approach—understanding function through structure—as the microscope.

With my microscope—a light microscope—I could see objects in color. Spectacular stains that work like dyes have been developed. They give objects brilliant colors; the stains differentiate different cells. Or they can show different functional elements of a cell or tissue, or the distribution of different proteins within cells. The electron microscope gives a higher level of resolution and shows things at a far greater magnification than a light microscope, but only in black and white. It permits the exploration of the very fine structure of cells. To examine a specimen with an electron microscope, you put an extremely thinly sliced piece of tissue in the path of the electron beam, and the beam projects an image of that section onto a screen. The final product is a photograph of that screen.

Because the magnification with an electron microscope is so great, you have a very small field of vision. To establish the orientation and position of the tissue or cell you can start at a low magnification, at a level similar to what you can see in a light microscope. Then, by increasing the magnification, the very fine details of organelles within a cell emerge. I began in electron microscopy long before computers were integrated into microscopes. The microscope itself was a large vacuum chamber, with an array of pumps to evacuate it. The electron beam was projected through the sample under study onto a screen, which had to be viewed in the dark. So the experience of using an electron microscope, in a darkened room with lots of noise from the vacuum pumps, felt very much as though you, yourself, were "in the microscope." I would spend hours in the microscope, scanning tissue, with a wonderful feeling of being inside the specimen.

The liver is a boring tissue to look at with the naked eye. Even the brain doesn't look particularly interesting without a way to see its internal structure. But when you observe these tissues with a light microscope you begin to see how impressive their structure really is. As you examine these tissues at higher and higher magnification, the beauty of the structure unfolds before your eyes, and you begin to understand how the structural elements of a tissue give rise to its function. We know from chemistry and physics that the structure of matter continues to be beautiful at ever higher resolutions. There is no end to the complexity of the structure and no end to its beauties. It is a bit like the experience of driving into the mountains. Last summer, my family and I drove into the Canadian Rockies from Calgary. From the outskirts of Calgary, the Rockies looked interesting and beautiful, but at a great distance, it was difficult to appreciate details. As we got increasingly closer, details emerged as the view of the rocks and forests expanded. The eye traveled over all the outcroppings and trees, and as we came closer still, the details of the sedimentation of the rocks and stands of different species of trees came into view, and then the details of the rock fissures, piles of stones, and of branches and leaves were revealed. The eye traveled, without tiring, across dimensions and emerging structures of seemingly infinite complexity.

After finishing college, I found a job in an electron microscopy lab almost by accident. One day after class, I asked a question of my cell biology professor, and he encouraged me to think about going into research. Before that conversation, I was sure I would go to medical school. Our conversation was crucial to my decision to pursue research instead. That conversation led to a job as an electron microscopist. Working in a microscopy lab was the most compelling experience I had yet encountered. I was drawn to the science, the studies of neuroendocrine areas of the brain that control an array of physiological processes throughout the body, and sharing that interest with the other people in the lab made a great difference. My graduate adviser at the National Institutes of Health, a man of enormous insight and enthusiasm, helped kindle my enthusiasm for neuroscience. He would come out of the electron microscope shouting, "Look at this. You've got to look at this." And we would study each other's latest findings and discuss their meanings with our colleagues in the lab.

The overarching goal of the lab was to understand the brain circuitry by which painful signals from the skin are relayed to the brain. The lab was part of a multidisciplinary group that included physiologists, pharmacologists, clinicians, and psychologists, all working to understand pain pathways and to design strategies to alleviate pain. The environment was vibrant. One of the great pleasures of science is that you do something you love and it also benefits others. Through my work I found joy and personal fulfillment; I also found the satisfaction of helping other people. And I was confirmed in a way of thinking that had been with me from childhood—my thinking with microscopes. It had never left me.

Not surprisingly, one of the first significant objects that my husband and I bought for our daughter was a dissecting microscope. Anything is interesting magnified. We kept the microscope in the kitchen so that we could look at something every day. We studied orange peels, leaves, flowers, bugs, newspaper print—anything and everything. You can easily spend an afternoon immersed in the intricacies of structure and how that structure leads to function.

These adventures in changes of scale are a way of thinking. We expand and contract time as well as space in our minds. I have spent a good part of my research life on the study of brain development. We study development through snapshots in time, but your mind links the snapshots together into a smooth progression. Changing scale in time and space is a wonderful tool for understanding. When I graduated to real microscopes, I understood the limitations of my first microscope, but to me, it had not been a toy. It had real optics. It magnified things. And it brought me a new route for understanding the living world.

Thinking with microscopes might seem to place you in a solitary frame of mind, but for me, the opposite was true. The work of science takes place in a community. The expectation for cutting-edge research, even as a graduate student, is to carry out a study that has never been done before, so, in a real way, you and your advisor are learning together. When I was working at the National Institutes of Health, it was early days for neurobiology, with people coming into the study of the brain from many different disciplines. I was persuaded to become a scientist not just by the science itself but also by the dynamics of doing science on a multidisciplinary team. Lab meetings brought together people from different fields, working on related problems. At every meeting, there would be a new idea that depended on the existence of the group. In my own lab's group meetings, I insisted that members share their plans, problems, and ideas to maximize our progress.

My scientific experience informs my current work as an academic leader. To understand any organization, you need to view it from many different levels at the same time and you need to understand its formal and informal patterns. I view MIT from the level of individuals—our students, faculty, staff, and alumni; from the level of the groups in which these individuals live and work; and from the level of the Institute in its national and international contexts. And I reach across and beyond the Institute to bring great minds to work on important challenges. Collaborative work, a "lab perspective," among different research groups and across departments, schools, and institutions, brings different viewpoints to bear on a problem and increases the opportunities for discovery. A "microscope perspective" informs through the discipline of considering situations at different scales and understanding their organizational structure. Things made transparent reveal their function. More than an instrument, the microscope becomes a way of thinking.

Susan Hockfield received a BA in Biology from the University of Rochester and a PhD in Anatomy and Neuroscience from Georgetown University at the School of Medicine. At Yale University, she served as the William Edward Gilbert Professor in the Department of Neurobiology, Dean of the Graduate School of Arts and Sciences, and Provost. She is currently President and Professor of Neuroscience at MIT.

What We Sense

PURPLE HAZE

Rosalind Picard

As a kid I liked school, but science was a turnoff. In my first school, science meant the teacher read to us about boring topics while we tried to sit still and be quiet. I remember when I picked out my favorite color notebook, yellow, for the class, thinking it might help me like science more, and thus help me survive. My next memory of science was third grade, this time in a new school, armed with a bright new yellow notebook. I honestly can't remember a thing about what we did that year except that I made a C, the worst grade I had ever received. I was off to a great start.

I managed better grades in science after that, but that was because I got very good at figuring out what the teachers wanted us to say and do. The instructions were almost always the same: read this, and take this test showing us that you read this. Science by recipe. You didn't have to understand anything to take these tests; you just had to be willing to find the words where they talked about the things the test was asking.

And then something changed. I think it was around fourth grade, when I had to do a science fair project and it could be on anything I wished. I chose penguins. I built a life-sized penguin and covered it with fur. My project on penguins won third prize at the school science fair. I think it was because of the great orange beak that my Mom and I sculpted out of putty after looking at a lot of books about penguins and experimenting with lots of sticky and stiff things in the kitchen until we got something that could be shaped just right. We also matched the color orange perfectly to the color in the big illustrated book of penguins. It was so much fun to build this amazing creature; for years I kept it in my room, staring at me, until one day it disappeared.

At the next year's science fair, my friend and I did a project on the pollution in the creek in our neighborhood. We finally got to really explore that creek. I learned how to use a little black-and-white portable camera to document what we saw. My dad pulled out his very expensive camera, a Rolleiflex, a big box-like fancy structure. When you looked down into the Rolleiflex you could see the scene you were shooting upside down. It was fun to be trusted with this special camera, but I had a hard time with the upside-down image and eventually went back to my cheap Instamatic.

Then, in sixth grade, I led the science class debate team arguing for Evolution against a team who argued for Creation. I showed an image with a sequence of monkeys turning into men, and everybody knew that Evolution was the only scientific argument there was. After the debate, the class voted to see who "won." I knew we would win because we were the only ones who used any scientific arguments, or so I thought. But we lost. The class voted for the Creationists, led by a cute popular girl with pigtails who played the guitar, had a great tan, and had a house with a pool. She spoke passionately about things that I didn't even want to hear. I couldn't believe it. I really had my nose out of joint after that, but maybe that was when I began to think people should know more science.

In high school I decided that I wanted to make As because this guy in my class used to taunt me in ways that made me so mad at him that I wanted to do better than he did. In high school, I had to take biology, chemistry, and physics. These sounded important. Also, the biology teacher was a coach, and he was really cute.

However, the first weeks of biology were once again reading and regurgitating. I found my Shakespeare class much more fun—the teacher pointed out all the bawdy parts of the plays and had us memorize passages from the greatest scenes. Our school also started an advanced-placement English class where we read and acted out scenes from great literature. Often these readings were mysterious, romantic, inspiring, and challenging. The stuff we had to memorize for science was none of that.

In chemistry I looked forward to actually mixing things and making things, but something unexplained always seemed to mess up our experiments. I wondered if it was the old glassware that always seemed to have some residue that wouldn't wash out. Once, we had a shiny-new, perfectly clear Pyrex flask, and I was thrilled because we were given a big procedure for an important experiment on which we were going to be spending a lot of time. I thought we finally had everything we needed to make it work. I took great care to follow the instructions perfectly, step by step, knowing that this time my experiment would have a chance of working if I did everything right. It was like following a recipe—you didn't have to know anything, you just had to follow the instructions and something wonderful would happen. I moved slowly, measuring carefully, checking the ingredients and amounts repeatedly, waiting just the right amounts of time, checking and double-checking my work. Finally, all that was left was to heat the beautiful solution in the lovely, sparkling-clean Pyrex flask. Finally when I attended to the heating of the solution, the heatproof flask cracked in half, and all the solution went into the flames. I felt like a failure when it came to chemistry.

I do have a vivid memory of one thing I learned in chemistry. My teacher was doing a class demonstration that required use of methyl violet indicator, a liquid dye that was a gorgeous deep shade of purple. During this demonstration, the rubber stopper got too deep into the mouth of the bottle and the teacher could not get it out. After several failed attempts, with class time running out, she decided to gently warm the container over the Bunsen burner so that the top would loosen and she could continue the demonstration. She looked very proud of her insight. We all watched as the stopper did indeed loosen—it shot up to the ceiling, and purple dye went all over the florescent lights, ceiling tiles, the teacher's white lab coat, her arms and face, her notes and books, the students in the front row, and more. I think this was the first time that class was adjourned and we wanted to stay! For days, maybe weeks, her skin still had purple spots. Years later when I visited the school, "Purple Haze" was still on the ceiling, and I felt delight as I fondly recalled that learning experience about how to remove a rubber stopper.

Through a special teacher, Fran Dubner, I learned about the Fernbank Science Center on the other side of Atlanta. I could skip out of afternoon classes to go a couple of times a week as long as I made up the work I missed in my classes. I would have to miss Typing (which everybody told me I should take since I "could always be a secretary") and Accounting (which everyone said was useful stuff to know). The work was trivial to make up and if I could look like I was skipping school then that would make me look pretty cool, and what teen girl didn't want that? I enrolled at Fernbank. I went with a couple of other students. Each time we went there, we would pile into my little-used car I had saved up to buy, and drive across town blaring the radio and singing like freshly uncaged birds. We'd often stop at a nearby Arby's, where kids sometimes got caught skipping school, and have a delicious cold Jamocha shake, leisurely, just so we could flaunt that we wouldn't get in trouble if we got caught while over there.

At Fernbank we met Mr. Tiller, who loved science and seemed really happy. He was young, tall, and kind of handsome, and I remember how he told us with pride that he was on his way to becoming Dr. Tiller. The first day he had us build a flip-flop circuit, which turned the lights on and off. This was really cool. I had never gotten to play with electronics before. He was willing to help us build anything we wanted. He let me take a Protoboard and function generator home with me. But the coolest thing of all was when I asked him about how holograms worked and he told me I could make them. He let me borrow a helium-neon laser, optics, film, instructions, and developing solutions and told me to give it a try. I quickly made friends with the girl on the high school newspaper who had a darkroom in her house. She also had a big dog who ran up and down the stairs making vibrations that ruined most of our holograms, but we soon managed to time our shots around the dog's schedule, and we produced some real holograms, completely on our own. It was exhilarating.

Dr. Tiller didn't tell us science was fun; he didn't have to. His sense of fun was contagious and could be seen in everything he did. His effect on me was as if he pulled out a stopper that had been used to plug up my curiosity, and then it could all start to come out. I didn't even know I was curious about so many things until he got us playing. I didn't know so many things were interesting, and that it was okay to just play with cool stuff. It was so much better than books.

Around this time I began skydiving, initially to overcome my fear of heights, and then I got hooked. I jumped because it was an incredible high without having to take the drugs that others in my school were turning to. Diving taught me many things—about the physics of acceleration, terminal velocity, and how it is really important not to look straight down at the ground right before you land. I loved "flying" where you could sail across the sky, in some cases attaining an almost 1:1 glide ratio. And I loved how movements in the sky did the opposite of what you expected—for example, reaching your arms forward to catch somebody would make you shoot backwards instead. I'll never forget the day I was driving home with both a laser in the passenger seat and a parachute in the back seat. I thought, I'm probably the only teenage girl in the world who has a laser and a parachute in her car today.

Later I was inspired to invite some of my pals from school to attend Tuesday night science talks at Georgia Tech. There we learned more about holography, the physics of vibrating strings, and more, but the thing I remember most was that the guys thought it was fun to go to science talks with girls.

I recently met Dr. Charles H. Townes, the man who invented the laser. I got to tell him how much this object had enchanted me. I told him how I was inspired by its power and by the teachers who loved what they did, who freed me to delight in asking questions and seeking answers. I fell in love with learning because I saw, and then felt, how much fun it could be. Over dinner with Townes and his wife, both nearly ninety, I got to see the spirit of playfulness with which both of them still spoke about science and the process of discovery—the very thing that had been missing in my early "cookbook" days as a student, the thing that had crept in little by little with the furry penguin and the purple haze and the cameras and holographs and vibrating strings, all of the cool and concrete things that turned me into the girl who loved to play with science, the lucky girl driving across town with a laser and a parachute in her car.

Rosalind Picard is Founder and Director of the Affective Computing Research Group at the MIT Media Lab and Codirector of the Things That Think Consortium. She holds a BS in Electrical Engineering from the Georgia Institute of Technology and an M.Eng and ScD in Electrical Engineering and Computer Science from MIT.

What We Model

STEPS

Moshe Safdie

Everything about Haifa, the city of my birth, seemed
to be about climbing steps. The city had originated on
the shores of the Haifa-Acre Bay. Some years before my
birth, the British built a modern port and the city began
climbing the slopes of Mount Carmel. I was born in a
Bauhaus-style apartment building abutting the Technion, the Institute of Technology, halfway up the mount.
By the time I turned ten we were elevated, in more than
one sense, and lived in a building up the hill. It was
a three-story building; the ground floor apartments
opened to gardens below and mid-level apartments had
side entrances. We lived on the top floor that we reached
through a bridge from the mountainside. We also owned
the roof of the building, which had an extraordinary
view of the bay and the Lebanon Mountains.

One hundred and seventy-three steps led up the
hill to the crest—Hadar Hacarmel, as it was known—
the center of my school and social life. In the morning
I would skip steps two at a time, rain or shine, to the
crest. From there I would either pick up a bike or a bus
to school. Returning in the afternoon, I perfected my
run, skipping four stairs at a time, timing myself, hoping to break my own record.

Agricultural stone terraces built hundreds of years
ago graduated the slopes. Steps abounded everywhere.
The land surrounding our building was nearly vacant.
As this was a time of austerity, we put the land to good

use. Earth was turned (often revealing centipedes or scorpions), and we planted tomatoes, cucumbers, and sweet peas. We built a hen house from leftover wood cases to supply eggs for the family.

One spring day, returning from school, I reported with great excitement that I could bring home a beehive. Everyone was short on sugar and the beehive promised honey. My parents were skeptical, but it arrived and was placed on the roof. It consisted of a basic box, a family of bees with its queen, and a series of empty frames, each with a base sheet upon which the bees could construct their house of wax. I watched with amazement as the bees went to work. Week by week, they constructed their house of perfect geometry, a series of hexagons, all seemingly identical, rising in orderly fashion, filling frame after frame. As the bees worked, I read profusely. I learned that the hexagonal cells were not all alike. Some were slightly bigger; in those, male eggs would be raised. Other cells, once filled with honey, would be sealed. I observed the queen, moving from cell to cell, laying her eggs. I watched for the wax moth—the curse of bees. I felt witness to the construction of an entire city, panel by panel, neighborhood by neighborhood. Different cells were specialized in function. Some were for the storage of food, others for raising future workers and queens, and some for the useless males, useless that is, except when their services were urgently required. This happened when an old queen was replaced by a new one and she needed her once-in-a-lifetime impregnation.

Even for an eleven-year-old, it was clear that this complex world was highly efficient. It exhibited a fitness to purpose that resulted in extraordinary beauty. I probably could not have articulated these words "fitness to purpose" until decades later, when at age twenty-two, by then an architect and apprentice in Louis Kahn's office, I was devouring D'Arcy Thompson's *On Growth and Form*. The memory of my bees made D'Arcy Thompson's

words more vivid. His studies of the lattice bone structure in the vulture's wing, the logarithmic growth pattern of the nautilus shell, the efficient space packing of
insects' habitats, were enriched by my having the beehive as a reference point.

In Haifa you could move up or down the hill via
steps everywhere, or you could move along the contours
of the sloped surface. Five hundred meters to the west
of my house was the world center of the Bahá'í religion,
the burial place of the founder of Bahá'í.

There relatively arid terraced land that had sustained a few olives, cypresses, and carob trees, was
transformed into a lush, green paradise garden. There
was a sudden transition across a wrought iron fence
that marked off a major estate of a hundred acres. Fash-
ioned in the Persian garden tradition, adapted to the
steep topography, there were monumental stairs made
of white marble that ascended up and down the hill on
its axis, that is, directly perpendicular to the slope, and
a series of gently meandering paths, accentuated by
long rows of cypress and palm trees. Some of the paths
were paved with loose chips of red tiles, others with
gray river pebbles, others yet with fine marble gravel.
There were colorful paths cutting through flowerbeds,
grass areas manicured to perfection, little round garden pavilions of white Corinthian columns and, always
on axis, bronze peacocks whose oxidized dark castings
were highlighted with gold leaf and silver. As I traversed
the paths, I had the feeling that I was sinking into a carpet, my body size diminishing, living within the rich ge-
ometries of the familiar Persian carpets that every room
in my parent's house possessed.

I abhorred Bible class. Five years later, exiled to
Canada, I would discover the Bible through an Angli-
can teacher at Westmont High School in Montreal when
I studied the book of Job and immersed myself in its
description of the paradoxes of life. But for now, I was
convinced that the paradise garden of the first book

of Genesis resembled my Bahá'í garden. As I walked through it, I would not have been surprised to come across Adam and Eve. There were certainly enough fig trees to supply the necessary attire.

In years to come, I had a recurring obsession in my work—I made buildings you could climb on, buildings that were gardens, gardens that were buildings, steps and gardens everywhere, even the title of my second book bears its trace, *For Everyone a Garden.*

By the time I was fourteen, my beehive had expanded. The one box had become two and then three, extending upwards in modular fashion. My family, unable to consume all the honey, shared it with relatives and neighbors. It was then that we received a new assignment in science class: to describe how energy is harnessed by making drawings and illustrations. Though I could draw with considerable ease, drawings seemed inadequate to describe what I had in mind. I joined with a friend and we decided to make a model. With a model, we could create a lake formed by a dam, show the water drop into turbines below, set windmills on the ridge, irrigate the terraces downhill, and so on. In the basement of a building that had been used as an air raid shelter during the Second World War, we found an old, unused door and used it as the base for our model. We purchased many pounds of clay to form hills, a lake, and valleys. We cut up little weeds to represent trees. We used dyes to color the landscape. We tried simulations by pouring water above and seeing it trickle downwards and we began searching for a pump that might keep the system going. We did all of this in a week of intensive work. We never attempted to move our model (nearly 4 × 7 feet in dimension) from our garden. By the time we had to bring it to school, it must have weighed a hundred pounds. Finally, with great effort, six of us tried to move it, ever anxious about cracking the clay. Eventually we hired a small truck, a transport arranged by proud parents, and our model was brought to school. Until then, I had been an indifferent student, at times flirting with expulsion. I had refused to do homework or comply with assignments. Now I was rising well above the call of duty.

I will never forget the excitement of making that first model. We had shaped the forms of the hills and streams. I felt like an extension of God's hands, shaping the landscape to my will. I felt a sense of what might come: conceiving and shaping buildings and landscape, a design process dependent on making models.

Only a year later, my life was disrupted by my parents' decision to immigrate to snowy, cold Montreal. I was on my way to a new world. Before we could leave for Canada, my family needed visas that could be issued only by the Consul General in Milan, Italy. The process took thirty days. To appease me and my brother and sister, all outraged by our enforced exile, my parents took us sightseeing: in Rome, there was the Forum, the Coliseum, the Vatican, and a place that transfixed me, Hadrian's Villa. Soon we were in Milan, where the roof of the Duomo, with its flying buttresses and fantasyland sculpture, became our playground. There were excursions to Lake Maggiore and Lake Como, to the floating island palaces in the lake, and the terraced gardens of Stresa. Within two years, as I completed high school year in Montreal and prepared to apply to university, there was no question in my mind. I would apply to the school of architecture.

Moshe Safdie has an international architectural practice designing a wide range of building types, including housing, arts, civic, and cultural buildings on four continents. After graduating from McGill University, he apprenticed with Louis I. Kahn in Philadelphia. He realized Habitat '67 at the 1967 World's Fair in Montreal and played a major role in the rebuilding of Jerusalem. He has taught architecture at Harvard, Yale, McGill, and Ben-Gurion universities.

What We Play

BLOCKS

Sarah Kuhn

In the slew of brochures that wash in on the daily
tide of the letter carrier's arrival, there is a Frank Lloyd
Wright catalog, where I find some gorgeous (and expensive) maple blocks. In dovetail maple boxes, the blocks
are elegant reproductions of the Froebel gifts—learning materials designed for young children by German
educator Friedrich Froebel, the creator of kindergarten.
They are beautiful. I must have them.

As a child, Frank Lloyd Wright had Froebel blocks,
and when he was a mature professional, Wright remembered them, saying that they were "in my fingers to this
day." Wright also said that the blocks taught him to see
in a particular way, and then to want to design. I am no
Frank Lloyd Wright, but on the day when the blocks arrived I knew what he meant.

The Froebel blocks are touchstones, connecting
me to the design studio I work in today and to days gone
by in the sunlit playroom of our old shingled house in
Berkeley. Light streams in through windows on three
sides of the room, illuminating the sun porch that has
become workshop, TV room, and flophouse for my sister and me. I am kneeling, my wooden blocks arrayed
before me on the worn Oriental rug.

I am deep in concentration, unaware of the passage
of time, or of the beautiful day outside the window. I
am immersed in the work of design and experiment,

constructing the walled city of my imagination. I am in the moment and in my body, fully engaged as I fashion tunnels, courtyards, and underground chambers from the plain wooden shapes. Some of the blocks are very large, 4 × 4s, clad in 1-inch yellowed and scuffed pine. I can feel the slightly rough sensation of the geometric pine blocks in my fingers, their edges rounded, their slight weight pressing against my hand as I grasp each in turn, then set it in its place.

The carpet makes a bumpy pattern on my knees as I kneel for long stretches, arranging each block, then sitting back on my heels to examine what I have done. The neighborhood I build is no larger than the spread of my arms, easily encompassed, ready to hand. I am flourishing in the sunlight like one of the giant plants that grows outside in my mother's lush California garden.

In one of my picture books, a character says, "I am the king of all I survey." That's how I feel, too, on this patch of rug in my playroom, but I have an extra edge—I *make* the world that I survey. It is mine, all mine, and I feel the deep joy of creating it. I am detached from the everyday. I am free to shape my castle, my neighborhood, as I will. I am free to change it as I want. Sometimes the blocks tumble, but I avert most catastrophes by building low structures that can't be destroyed by a single mishap. In my city, my people obey the same rules that govern the outside world—they cannot walk through walls, they need structures that are stable and safe. Most of the time though, they act according to their own logic, which is also mine. But mostly, the people are not there yet, and often they never arrive. For me the best part is the preparation, the planning, and the design. Everything is literally within my grasp, as I finger the pieces of wood. A walkway that leads to a ladder that leads to a higher walkway. A long and winding tunnel with boats in it. A large underground room.

My bedroom and the playroom are upstairs and my father's study is downstairs. Unlike my light-filled play areas, his workroom seems a dark lair, in a darker part of the house and with curtains partly drawn. I see him sitting in the middle of the room, back to the door, erect at his typing table. His typewriter is a big, heavy manual and his large fingers are strong from the thousands of hours he has spent over the years pushing down those heavy keys. He has a cigarette dangling from his lips as he types. Its smoke makes a thin curl that winds its way upward toward the ceiling. By dinnertime the room is filled with haze, from hours of thought and smoking, so that when I go to call him for dinner I see him as if through a mist. But perhaps this only makes his lair the more enchanting. I see his typewriter as the alchemical vessel that commands his attention. What is he doing all those hours punching away at that machine? Many years later, my mother's uncle told me of his time in the Far East, and taught me some of the pidgin English he had learned. Piano was "bakkus (box) you fight 'em teeths." The typewriter had teeth, too, but my father was not fighting with it; he was making the keys fly and letters appear on the page. I begged him to let me try it.

I remember the day I am finally allowed to try my father's typewriter. I am about five, just starting kindergarten, and my father ushers me into his study. Everything is hushed, and I am impressed by the solemnity of the moment. At last I will get to do what I have seen my father doing. I raise my hands above the keyboard, fingers curving downward, and bring my hands down hard toward the keys, just as I have seen my father do. Disaster! Instead of letters in a neat row on the page, I have in front of me a multi-key accident, a tangled mass of metal. I am sure I have done exactly what I had seen my father do; yet the result is completely different. My sense of mastery and that I know what is going on in the world around me, is shaken to its foundation. How can things be so much more complicated than they seem? I return to the playroom to lick my wounds. My blocks give me solace, obeying my every command, even when the rest of the world does not.

After my brother is born, my parents convert part of the attic into bedrooms for my sister and me. Having my own bedroom, my private universe for my construction, is a new experience. The scope of my ambition now expands to fill the room; the bed and the floor become part of the action. My grandfather builds me a table in the form of a giraffe, and I incorporate it, too, into my constructions. Usually my bed is a raft, a safe harbor in the world of danger. Anything touching my bed is part of the raft, keeping me and my stuffed animals safe and dry. The giraffe is right up against the bed, and fingers of wooden walkway, formed with my trusty wooden blocks, fan out into the room to allow me to venture away from the bed. Back in the playroom, I would have had to imagine myself an inch high to live in my blocks compound; in my bedroom, I can be my own size. I inhabit my construction world; I am both designer and client.

In my bedroom, the new blue carpeting is a hostile sea, its unplumbed depths harboring countless marauding sharks, the bogeymen of my childhood. (We live only a few miles from Alcatraz, the notorious maximum security prison island, and I shiver at stories of would-be escapees who come to a bad end.) As I extend my construction, I extend my world of safety.

The summer after seventh grade we move across the country, to the terra incognita of New Jersey. My wooden blocks are among the many things we give away when we move. But, in my mind, I had left them behind long ago. Wasn't I almost grown up, ready to put aside these childish things? Didn't I have much more important work to do? Yet as I thought of career, my emerging interests led me back to the blocks. It is my senior year in college. I am majoring in philosophy and sociology, and I have finally settled on the topic for my senior thesis: toys. I call the finished work "The Objects of Play" and reflected on questions like, What is a toy? What makes something toy rather than "real"? What is play and how do physical objects figure into it?

In graduate school, I become interested in the impact of information technology in the workplace, which involves me, of necessity, in the world of computers. When I begin my work, computer interfaces are unattractive and cumbersome objects—line or dot-matrix printers clank loudly; monochrome text of obscure code glows green on cathode-ray tubes. But within a few years, computing discovers graphics. The visual blossoms in the dry symbolic world. Growing up, I identify with my mother's interest in design, particularly modern design, and my father's friendship with the designer Charles Eames. During college I spend a wonderful summer working at the Eames office in Venice, California; after college, I put these interests aside. They seem frivolous, not socially useful; there is more important work to do in the world. But now, looking back over several decades, my artistic interests are more of a piece with how I work with software—design after all. My early interests seem fundamental, a phantom limb become real again.

Sarah Kuhn is Associate Professor of Regional Economic and Social Development at the University of Massachusetts, Lowell, and founder of the Laboratory for Interdisciplinary Design. Her current work is at the intersection of socially responsible design, the integration of technical education with the social sciences, and service-learning, which is a method of teaching, learning, and reflecting that combines academic classroom curriculum with community service.

What We Build

RADIO

Donald Norman

How did my career begin? Ah, through a wonderful, wooden piece of furniture.

It was a radio, perhaps two feet tall, one foot wide. In a wooden case, of course, and in my memory, it has a rounded top and a large round dial for tuning the AM band (FM didn't really exist as a practical broadcast medium in those days). I remember a large circular opening at the top front of the radio for the speaker, covered with a grill cloth. But the real joy was inside, visible only through the rear. There within were arcades of circuitry illuminated, cathedral-like, by the glowing filaments of the vacuum tubes. How many? Memory fails me, but six or eight feels right. Vacuum tubes—diodes, triodes, pentodes, or even two triodes in a single glass envelope.

I loved the insides of the radio. I can remember the undersides, a mesh of thick wires running this way and that, covered with dust and cobwebs, connected at junctions with nice dull solder balls, with multiple large cylinders connected to them. Smaller, compact cylinders with multiple color bands—these were the resistors, and eventually I learned to decode their values by learning mnemonics. Larger cylinders were condensers (now known as capacitors). There were even coils of wire—inductors.

But the real mystery is how it worked. I was quite used to mechanical devices and had long irritated my parents by taking household objects apart so that I could understand how they worked (and invariably, failing to put them back together again). But the radio offered no clues to its operations. There was no obvious relationship between the radio's moving parts and the functions they accomplished. In the radio, the moving parts simply added to the mystery. The tuning knob rotated a pulley, which had a piece of string stretched around it that also went around a pulley attached to a variable capacitor: a strange device with two sets of parallel plates, one set of plates stationary, the other rotating as the knob rotated, its plates interleaved with the stationary ones. This is how one tuned, but it made no sense—how did that select stations?

This is how my lifelong quest to understand the workings of invisible things began, with that radio. I discovered a set of books, written for people just like me, that led me down the path of understanding electricity, basic electronics, and the workings of radios. I discovered the world of Radio Amateurs, active in those days with periodicals, an association, and clubs. While still in junior high school, I became a certified radio amateur and built my own transmitter and receiver. Morse code only, of course.

By then, my love of radios had reached an advanced stage of lust. I pored over catalogs, dreaming of ever more powerful receivers and transmitters, ever taller and more sensitive antennas. As I did so, I was learning about electronic mechanisms, the laws of physics, and mathematical reasoning. But just as my ability to take things apart was not matched by an ability to put them back together again, my skills with the clean, accurate, pristine mathematics of circuit design were not matched by skills within the prosaic world of screws, drills, relay racks, wire cutters, and hot, messy soldering tools. The things I built never quite lived up to expectations. My circuits eventually worked, but accompanied by burnt fingers, bruises and cuts. In retrospect, this should have signaled that I was a theoretician, not a builder. At the time, all I got out of it was frustration.

But the radio transformed my life. I finally had focus: to understand the hidden mechanisms of electronics. This desire motivated high school interest and university degrees in Electrical Engineering. Even after my master's degree, when I changed disciplines to pursue a doctorate in psychology, my goal remained the same. I still needed to understand the mysteries of hidden, invisible mechanisms, but now my focus was on the human mind rather than the electronic circuit.

There is a magic to the invisible world of electronics, but electronics is still about physical things: electrons and electrical circuits. There is even more intrigue to the less tangible world of information. This is the world of computers and computation, and today, of thinking about the human mind. People, of course, are far more than computational objects that process information. We live in a complex, tangible world, where our activities are colored by social interaction, culture, and emotions. But the path to understanding even these aspects of human life has the same excitement and allure for me that I felt in my original quest to understand the magic of the radio. In other cases, whatever is going on is not magic, not forever to be a mystery. The exploration of mystery is what science is about, that along with the promise that answers can be found.

Donald Norman, with a long career as a cognitive scientist in the close psychological study of everyday things, codirects a joint MBA and engineering program in design and operations at the business and engineering schools at Northwestern University, writes books, and serves on boards such as the Korea Advanced Institute of Science and Technology and the editorial advisory board of the Encyclopaedia Britannica. He is a principal of his technology consulting company, the Nielsen Norman Group.

What We Sort

VENUS PARADISE COLORING SET

Donald Ingber

When I was a child, the only way my mother could get
me to the dentist was to bribe me with the promise of a
Venus Paradise Pencil by Number Coloring Set. It consisted of a cardboard box filled with six colored pencils
emblazoned with silver numbers, a red plastic pencil
sharpener, and a sheaf of black-and-white drawings of
sentimental scenes of American life, each numbered for
easy coloring.

The box itself was exciting. "Venus" was spelled
out boldly; "Paradise" was written in a style reminiscent
of a Disney film advertisement; "Coloring" was larger
than the other words, in capital letters, each a different
color of the rainbow. The package cover, a product of
mid-1960s advertising, included slogans such as "NO
WATER / NO BRUSH / NO MESS"; "IT'S FUN / IT'S EASY";
and the main selling point: "A BRILLIANT COLORED PIC-
TURE EVERY TIME." The boxtop indicated the type of
scenes that were depicted on the enclosed paper sheets
("PLAYTIME," "SPORTS IN ACTION," "AMERICA IN SPACE").
Finally, the side of the box detailed its exact contents: 5
Numbered Pictures, 6 Colored Pencils, 1 Pencil Sharpener. I was fascinated by this contrast between the hype
of the message and the precision in this economy of
description.

When I removed the cover of the box, there was a
folded piece of red cardboard that formed an elevated
rectangular rib extending across the box, partially cov-

ering the paper sheets (each pre-titled) and the bottom of the box below. The pencils were positioned on this rib, each parallel to its neighbor, as if suspended in mid-air. This was accomplished by slipping each pencil through two holes that were punched out of opposite vertical sides of the rectangular rib. The pencil sharpener sat snugly beside the pencils in its own rectangular hole cut out of the upper surface of the folded cardboard. The raised pencil holder area was labeled "VENUS PARA-DISE COLOR SELECTOR" and there were other slogans as well, each optimistic and conveying a can-do mental-ity: "VENUS PARADISE LEADS ARE STRONGER-SMOOTHER WITH THE COLOR BRILLIANCE OF OILS" and the ultimate promise: "First time . . . Every time . . . A BRILLIANT COLORED PICTURE WITH A PROFESSIONAL TOUCH."

The serious business of coloring by number began when I chose a pre-numbered picture to color. The sketched drawings had enclosed and numbered spaces that each corresponded to a pencil number. I was im-mediately confronted with many troubling questions: Should I press hard with a dull point and fill in each box with vibrant color, or should I sharpen the point of the pencil until it glistened and then gently graze the sur-face of the paper? If I took the gentle approach, should I carefully position each scratch of the pencil parallel to its neighbor, neatly filling in the polygonal space like a French pastel drawing, or should I scratch wildly like a cat?

But there were deeper questions: should I follow orders and use pencils with the same numbers as those printed on the sheet, or should I see what happens if I choose my own colors? Should I fill in the spaces from top to bottom, left to right, starting at the center, or work at random? Finally, should I abandon all caution and go crazy like my friends' younger sisters and broth-ers by ignoring the lines between the numbered spaces and along the outer borders of the scene? The choices were stark and they were mine: play by the rules or break out and see what happens. So, at first glance, the Venus Paradise Coloring Set seemed to demand constraint, with all creativity stifled by rigid instructions that dictated what pencil would be applied to what exact spot on paper, but in fact, the set was more like a Rorschach test, a projective screen onto which each user could see and do something different.

I grew up on Long Island, about fifteen miles east of Manhattan, in one of Manhattan's first suburban communities. My town itself was much like a Pencil by Number drawing titled "East Meadow, New York." It was composed of thousands of contiguous polygonal plots of land, each filled in with nearly identical cookie-cutter homes, each differing from its neighbor only by its characteristic number and color. It was life in the middle of the middle class, in the middle of the first suburban sprawl. My parents had never completed college and had worked hard since they were teenagers. They had no apparent appreciation for art, and I have no memory of ever being taken to a gallery or museum. I do remember that when my mother went with the Women's Sisterhood from our temple to visit the Museum of Modern Art in New York City, she returned with a gift for me. It was a postcard depicting a Mondrian painting composed of a few overlapping vertical and horizontal black lines on a white background that mapped out a bare rectangular grid. My mother laughed and said that it was so simple that either of us could easily have done it ourselves. But this painting was worth a million dollars!

It was twenty years later on a traveling fellowship in England (pursuing interests in cancer research, theater, and art) that I came to my own understanding of Mondrian's painting. Mondrian essentially ran the Venus Paradise Pencil by Number Coloring process in reverse—and with paint. He started by painting conventional full color, realistic depictions of churches, seascapes and trees, and then progressively removed one bit at a time—first the colors, then the lines—until he portrayed what he saw as the essence of the image. His vertical and horizontal framework was analogous to the scenes in my coloring set after all.

When I was a child, I did not understand art but was drawn to it. Patterns and forms caught my attention and made my heart pulse more than sounds or music. I had a keen ability to see how things worked. To my mother's dismay, I routinely disassembled and reassembled bicycles and broken televisions. It was perhaps such interests that caused my friends' parents to give me science toys for my birthday presents every year, but I remember that I ignored these gifts. I would put them in the back of the bottom drawer of the old chest that I used for toys. Instead, I would pull out my Erector Set, Tinkertoys, Lincoln Logs, or Venus Paradise pencils and set out to make something all my own. It wasn't that science didn't interest me, but these kits all seemed to require that I follow rigid rules in order for their experiments to be successful. This was not as interesting as exploring what I could build without rules. I loved to see whether or not structures would hold their shapes when I released my hand, or what pictures I could draw by filling in numbered spaces with my vibrant colored pencils, often ignoring the preprinted numbers.

The Venus Paradise Coloring Set had another attraction. I could create beautifully colored scenes and clearly recognizable images without having to take an art class, do homework, or receive any input or instruction from an adult. This felt like a great accomplishment. And the set also taught me that limitations can be circumvented. I am colorblind. The numbered pencils and the drawing templates allowed me (when I so wished) to circumvent the terror I experienced when a teacher asked me, for example, to paint something green, and I had no idea which paint cup to select.

I gained much more from the coloring set. After coloring in multiple scattered spaces, I was always elated when I penciled in that key space that caused all the other colored tiles to merge into a single coherent image. The moment always came suddenly, a surprise I learned to anticipate with great expectation. It was in this way that I came to understand the power of the gestalt, that the whole is greater than the sum of its parts, and that the overall arrangement of the parts can be as important as the properties of these components.

Of course, the coloring set had its limitations. By deconstructing the scene and reducing it to a spare framework, its full richness was lost. No matter how perfectly I colored in the Pencil-by-Number diagrams, the final product was never really right. It was not engaging, not art. Something was missing. Although I succeeded academically based on my gifts in mathematics and science, I sought out opportunities to better understand what was missing in those pencil drawings. I was frustrated when I entered Yale College because it was nearly impossible for a science major to take studio art. I nearly gave up after an unsuccessful midnight interview for entry into a painting course. But I became intrigued when I saw students walking around campus carrying polyhedral sculptures made of folded cardboard that were very similar in form to the viruses that I was studying in my science class on molecular biophysics.

The students were taking a sculpture course called "Three Dimensional Design." This fascinated me because in my science class, I had learned that the three dimensional shapes of molecules governed their functions. The double helix of DNA is the classic example in which the complementary shapes of two strands of DNA must match perfectly in order for them to be zipped together to form its single molecule, a molecule that some view to be the very essence of life. And the enzyme molecules that catalyze biochemical reactions physically interact with their substrates in the manner of locks fitting keys; the chemical conversion that takes place results from the enzyme's ability to change its shape, thereby mechanically deforming or breaking its key-like substrate.

I found a way to talk myself into this sculpture course, and it was there that I had my first "Aha Moment," one that launched me on a path that I follow to this day. Erwin Hauer, who taught the class, was lecturing on a strange structure he called a tensegrity sculpture. It was composed of six wood dowels that appeared to be suspended in mid air by a tensed network of multiple elastic cords. I learned that the mechanical force that balanced between the pull of the strings and push of the struts placed the whole skeletal framework in a state of isometric tension. The force stabilized the shape of the skeleton of stick and string, much the way the mechanical forces balanced between muscles and bones stabilize the shape of our bodies. It was the inventor of the geodesic dome, Buckminster Fuller, who had named this building system that depends on tensional integrity. The stick-and-cable structures that conveyed the essential principles of tensegrity were conceived and constructed by his student, the artist Kenneth Snelson.

When Hauer pressed down on this spherical framework, it spontaneously flattened, and when he released his hand, the structure rounded and jumped up in the air due to the release of energy stored in the stretched cables. The movement and shape transformation transfixed me. I had seen nearly the same behavior only a day before. Except in that case, I was observing cancer cells I had just learned how to culture in a research laboratory across campus at the medical school.

Living cells are round when they are removed from our tissues, but they spread and flatten when they anchor to the surface of a culture dish. Once the cells divide and multiply to cover the substrate, they must be removed and passed to multiple new dishes in order to continually expand the population. This is accomplished by adding an enzyme to the adherent cells that clips their molecular anchors. When this happens, the flat cells immediately round and leap up into the culture medium, just like Hauer's tensegrity structure.

I had my "Aha Moment" in the mid-1970s, at the same time that scientists were first discovering that all cells have an internal molecular framework, or cytoskeleton, similar to the one that generates tension in our muscle cells. My pencil-by-number mind filled in the spaces of the tensed three-dimensional framework that I watched transforming back and forth between sphere and pancake beneath Hauer's hands. I saw that cells must be built this way. Cells must be tensegrity structures.

When I first described my ideas to senior scientists, they did not respond well. Most scientists still thought of living cells as little bits of viscous protoplasm surrounded by elastic membranes. Most scientists continued to be critical of my ideas, no matter what my academic achievements. I responded to skepticism with patience (Venus Paradise artists need this too!), and after many years, I have been able to draw a picture of how life works at the cell level that most scientists now find convincing.

Pursuit of tensegrity has led members of my laboratory in unanticipated directions. Mechanical forces alter cell structure, and we have found that the cytoskeleton that governs the shape of cells is as important for biological regulation as are chemicals and hormones. This art-inspired insight has implications for cancer research, embryological development, and even understanding the sensation of gravity. It has contributed to the development of anti-cancer therapies and nanotechnologies. Most amazing to me, it has led to a new way of thinking about the origins of life on this planet.

I believe that my scientific contribution drew crucially on my Venus Paradise training. At the school of Venus Paradise, I learned that there is structure to pattern and that function follows form, rather than the other way around. From Venus Paradise I learned that there can be simplicity in complexity, and that art and science are one and the same. If a visually rich image can be conveyed with simple rules and translated into what is essentially a black-and-white engineering diagram, then it might be possible to discover guiding principles that govern how more highly complex structures integrate form and function, including living cells and tissues.

Now my life is at the art-science interface. My scientific vision is based on art and architecture. There is an aesthetic in the scientific images we create in pursuit of life's inner secrets. My research group has produced large prints of microscopic images of the cytoskeleton, brilliantly colored with fluorescent reds and greens. These prints have been included in museum art exhibitions and published in international art and architecture journals. I regularly lecture to artists as well as scientists, to architects as well as engineers. I teach Harvard College students about life at the interfaces of art, science, engineering, physics, and medicine.

My Venus Paradise Coloring Set conveyed to me that everything has underlying structure, so even life can have architecture. It was creativity in a box, all enclosed in one neat little package containing 5 Numbered Pictures, 6 Colored Pencils, and 1 Pencil Sharpener. Ironically, its constraining diagrams and rules helped me break out—from my town, from rigid disciplinary definitions, and from ways of thinking that divide art, science, and technology.

Donald Ingber is the Judah Folkman Professor of Vascular Biology in the Departments of Pathology and Surgery at Harvard Medical School and Children's Hospital, Boston. He received his BA, MA, MPhil, MD, and PhD from Yale University, and carries out research at the interface of biology, medicine, engineering, and the physical sciences, focusing on how living cells and tissues form and function.

What We Program

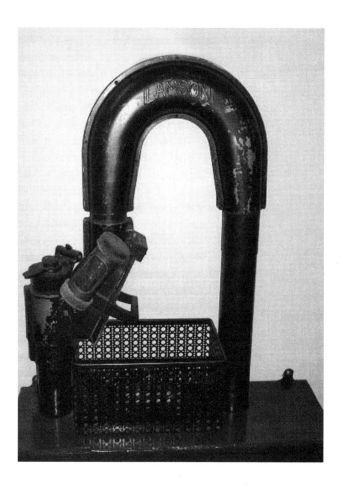

VACUUMS

Alan Kay

My fourth grade teacher, Miss Mary Quirk, was different from the others.[1] The other teachers wanted us to read One Book, the textbook that they used as the final authority for all opinions, the answer to all questions. The teachers got angry at questions that challenged the One Book. What I learned from those One Books was to avoid them and retreat to many books. The One Books taught me that if I wanted to learn something, I could do it myself.

But Miss Quirk was different from the others. In her classroom, there was an old dining table toward the back on the right-hand side that was completely covered with various kinds of junk: not only books but tools, wires, gears, and batteries.

Miss Quirk never mentioned this table. Eventually, I started to poke around to see what was on it. Predictably, I was drawn to the books. One of them was about electricity. One afternoon during an English class I set up my English book with the smaller electricity book behind it, and the large dry cell battery, nail, wire, and paper clips behind that. I wound the bell wire around the nail as it showed in the book, connected the ends of the wire to the battery, and found that the nail would now attract and hold the paper clips!

I let out a shriek: "It works!" The class stopped. I hunched down expecting some form of punishment—in school, this had often happened to me after such transgressions. But Miss Quirk did nothing of the kind. She stopped the class and asked me, "How did you do that?" I explained about the electricity book and showed my electromagnet holding the paper clips. She said: "That's great! What else is in the book?" I showed her that the next project was to make a telegraph with the electromagnet. She asked if others in the class were interested in this, and some were. So she said, "Okay, later this afternoon we'll have time for projects and you all can work together to do the next thing in the book." And that's just what happened.

And it happened many, many times. Children would find stuff on the table and want to make things. Miss Quirk would ask them to show the objects of their enthusiasms to the rest of the class to check out who else might be interested in working with them. Pretty soon about half our class time was devoted to these self-chosen projects. We started showing up earlier and earlier for school in the hope we could spend more time on them. Miss Quirk was always there. We could never beat her to class.

Most of my ideas about how elementary school education should be done are drawn from the way Miss Quirk ran her classroom. She took projects that interested children and integrated real mathematics, science, and art into them. That was her curriculum.

Years later when I lucked into a terrific graduate program at the University of Utah, my first thought was that this was just like fourth grade. And then I realized that Mary Quirk had made fourth grade just like a great graduate school. This is a critical insight. Children are in the same state of *not knowing* as research scientists. They need to go through many of the same processes of discovery to make new ideas their own. Because discovery is difficult, children have to be given scaffolding for their ideas. They need close encounters with rich materials; they need a careful yet invisible sequencing of objects that will enable them to make the final leaps themselves. This was the genius of Mary Quirk. We never found out what she knew. She was focused on what *we* knew and could find out.

Emboldened by Miss Quirk, I began to find things out. One day, I noticed that a local department store had a pneumatic tube system for moving receipts and money from counters to the cashier's office. I tried to figure out how it worked and asked the clerks about it. They all said the same thing. "Vacuum," they said, "vacuum sucks the canisters, just like your mom's vacuum cleaner." "But how does it work?" I asked. "Vacuum," they said, "vacuum does it all." This, I discovered, was what most adults call "an explanation."

So I took apart my Mom's Hoover vacuum cleaner to find out how it worked. There was an electric motor in there, which I had expected, but the only other thing in there was a fan. How could a fan produce a vacuum, and how could it suck?

We had a room fan and I looked at it more closely. I knew that it worked like the propeller of an airplane, but I'd never thought about how those worked. I picked up a board and moved it. This moved air just fine. So the blades of the propeller and the fan were just boards that the motor kept moving to push air.

But what about the vacuum? I found that a sheet of paper would stick to the back of the fan. But why? I had heard that air was made up of particles too small to be seen. So when you got a gust of breeze by moving a board, you were knocking little particles one way and not the other, like rowing an oar. But where did the sucking of the fan and the vacuum cleaner come from? Why did the paper stick?

Suddenly it occurred to me that air particles must always be moving—very quickly and bumping into each other. When fan blades move air particles away from the fan this put fewer particles near the fan. Now already moving particles would move toward the fan, their path cleared. These particles didn't "know" about the fan, but they appeared to. The "suck" of the vacuum cleaner was not a suck at all. Things went into the vacuum cleaner because they were being "blown in" by air particles moving into the fan, unopposed by the usual pressure of air particles inside the fan.

When my father came home that evening I told him what I'd found out: "The air particles must be moving at least a hundred miles an hour!" He looked in his physics book for a formula to compute the speed of air molecules at different temperatures. It turned out that at room temperature ordinary air molecules were moving much faster than I had guessed: more like 1,500 miles an hour!

An old mechanical fortune-teller in a Thornton Wilder play says: "I tell the future: nothing easier, but who can tell the past?" The future, vague and approximate, is somewhat easy to tell; the past is messy. But this episode with the vacuum cleaner, the fan, my father, and his physics book is what I remember to be the first time I ever thought like a scientist, resisting common sense long enough to get at the heart of the "Why?"

Alan Kay is one of the earliest pioneers of personal computing (including the idea of the laptop), GUIs (overlapping window interface that is ubiquitous today), and object-oriented programming. Many of these inventions were inspired by the possibility of helping children learn to think better than most adults today.

Objects in Mind

GEARS

Seymour Papert

Before I was two years old, I developed an intense involvement with automobiles.[1] The names of car parts made up a substantial portion of my vocabulary: I was particularly proud of knowing about the parts of the transmission system, the gearbox, and most especially the differential. It was, of course, many years later before I understood how gears work; but once I did, playing with gears became a favorite pastime. I loved rotating circular objects against one another in gear-like motions and, naturally, my first "erector set" project was a crude gear system.

I became adept at turning wheels in my head and at making chains of cause and effect: "This one turns this way, so that must turn that way so. . . ." I found particular pleasure in such systems as the differential gear, which does not follow a simple linear chain of causality since the motion in the transmission shaft can be distributed in many different ways to the two wheels depending on what resistance they encounter. I remember quite vividly my excitement at discovering that a system could be lawful and completely comprehensible without being rigidly deterministic.

I believe that working with differentials did more for my mathematical development than anything I was taught in elementary school. Gears, serving as models, carried many otherwise abstract ideas into my head. I clearly remember two examples from school math. I saw multiplication tables as gears, and my first brush with equations in two variables (e.g., $3x + 4y = 10$) immediately evoked the differential. By the time I had made a mental gear model of the relation between x and y, figuring how many teeth each gear needed, the equation had become a comfortable friend.

Many years later, when I read Piaget, this incident served me as a model for his notion of assimilation, except I was immediately struck by the fact that his discussion does not do full justice to his own idea. He talks almost entirely about cognitive aspects of assimilation. But there is also an affective component. Assimilating equations to gears certainly is a powerful way to bring old knowledge to bear on a new object. But it does more as well. I am sure that such assimilations helped to endow mathematics, for me, with a positive affective tone that can be traced back to my infantile experiences with cars. I believe Piaget really agrees. As I came to know him personally I understood that his neglect of the affective comes more from a modest sense that little is known about it than from an arrogant sense of its irrelevance. But let me return to my childhood.

One day I was surprised to discover that some adults—even *most* adults—did not understand or even care about the magic of the gears. I no longer think much about gears, but I have never turned away from the questions that started with that discovery: How could what was so simple for me be incomprehensible to other people? My proud father suggested "being clever" as an explanation. But I was painfully aware that some people who could not understand the differential could easily do things I found much more difficult. Slowly I began to formulate what I still consider the fundamental fact about learning: Anything is easy if you can assimilate it to your collection of models. If you can't, anything can be painfully difficult. Here too I was developing a way of thinking that would be resonant with Piaget's. *The understanding of learning must be genetic.*

It must refer to the genesis of knowledge. What an individual can learn, and how he learns it, depends on what models he has available. This raises, recursively, the question of how he learned these models. Thus the "laws of learning" must be about how intellectual structures grow out of one another and about how, in the process, they acquire both logical and emotional form.

I work on an applied genetic epistemology expanded beyond Piaget's cognitive emphasis to include a concern with the affective . . . a new perspective for education research focused on creating the conditions under which intellectual models will take root. . . . In doing so I find myself frequently reminded of several aspects of my encounter with the differential gear. First, I remember that no one told me to learn about differential gears. Second, I remember that there was *feeling, love,* as well as understanding in my relationship with gears. Third, I remember that my first encounter with them was in my second year. If any "scientific" educational psychologist had tried to "measure" the effects of this encounter, he would probably have failed. It had profound consequences but, I conjecture, only very many years later. A "pre- and post-" test at age two would have missed them.

Piaget's work gave me a new framework for looking at the gears of my childhood. The gear can be used to illustrate many powerful "advanced" mathematical ideas, such as groups or relative motion. But it does more than this. As well as connecting with the formal knowledge of mathematics, it also connects with the "body knowledge," the sensorimotor schemata of a child. You can *be* the gear, you can understand how it turns by projecting yourself into its place and turning with it. It is this double relationship—both abstract and sensory—that gives the gear the power to carry powerful mathematics into the mind. The gear acts here as a *transitional object.*

A modern-day Montessori might propose, if convinced by my story, to create a gear set for children. Thus every child might have the experience I had. But to hope for this would be to miss the essence of the story. *I fell in love with the gears.* This is something that cannot be reduced to purely "cognitive" terms. Something very personal happened, and one cannot assume that it would be repeated for other children in exactly the same form.

My thesis could be summarized as: What the gears cannot do the computer might. The computer is the Proteus of machines. Its essence is its universality, its power to simulate. Because it can take on a thousand forms and can serve a thousand functions, it can appeal to a thousand tastes. My own work over decades is to turn computers into instruments flexible enough so that many children can each create for themselves something like what the gears were for me.

Seymour Papert, one of the early pioneers of artificial intelligence and one of the main developers of the Logo computer language, co-founded the MIT Artificial Intelligence Laboratory, founded the MIT Media Lab's Learning and Epistemology Group, and advised the One Laptop per Child project, a nonprofit association dedicated to developing and providing children around the world with a $100 laptop.

EPILOGUE: WHAT INSPIRES?

Sherry Turkle

We cannot know whether we stand before a child who will use objects as a path to science. And we cannot know to which objects a child will be drawn. The memoirs of this collection teach the importance of acknowledging and accommodating such uncertainties. A one-kind-fits-all curriculum is likely to take children away from the objects that compel them. Insisting on uniformity, we might miss a child who makes Cs and Ds in math and science but develops an abiding love for computers because of his connection with LEGOs. We could miss a child who doesn't think of herself as a science student even as she absorbs everything she can learn from lasers and "purple haze" chemistry. We might not count as learning the lessons that come with braiding a pony's tail, stoking a wood stove, or baking a meringue.

In *Mindstorms: Children, Computers, and Powerful Ideas,* Seymour Papert writes of falling in love with the gears of a toy car that his father gave him when he was two. Papert fell in love with the gears and through them fell in love with science. The gears inspired, but Papert makes the point that if anyone had tried to test him to determine what was happening as his curiosity was expanding, they would have found nothing to measure.

The story of Papert's gears makes plain that finding nothing to measure does not mean that nothing is going on. Too often, if we can't formulate a test, we give up on a method or we give up on a child. The voices in this collection remind us that at these moments, at the limit of measurement, we can turn directly to a child and put our deeper intelligence to work. It can be a moment when we listen, when we learn what inspires, a moment of discovery for a parent or a teacher.

As we face a national crisis in science education, we are drawn to the next new thing, and these days, that usually means digital media. But we can also look to things that have worked in the past—and one thing that has reliably inspired scientific curiosity has been object passions.

The essays in this collection bring us to an excellent vantage point for thinking about science education because they directly pose the question, "What inspires a young scientific mind?" Objects bring us to details. How are inspiring objects chosen? How are they mastered? What states of mind do they encourage? Papert, when he considered these questions, focused on the work of his mentor, Jean Piaget. There, the emphasis was on the cognitive. But Papert also said that he fell in love with the gears, an observation that brings us to what the psychoanalytic tradition has to contribute to thinking about young people and the awakening of science. Psychoanalysis works by investigating forms of love and how they shape the heart and mind. It has much to say about love of objects and how they can spark new learning, about the space where science can happen, and the relationship between the scientist and his or her materials of thought.

In 1975, the French psychoanalyst Jacques Lacan came to MIT to try out some new ideas about objects and scientific inspiration. Lacan had a specific object in mind: knots. For Lacan, playing with knots wears down our resistance to the intimate relationship between topology and the body. Manipulating and perforating little circles of string frees us up to fully engage with the mathematics in knots. Before he gave his talk, Lacan painstakingly drew knots in four colors on the green chalkboards of his seminar room.[1] Most of his audience assumed that Lacan's knots described a model of mind, but Lacan was trying to say something different—that knot-play is a critical element in the emergence of insight about science, in the same sense that psychoanalytic insight grows out of a lived relationship between patient and analyst.

Lacan focused attention on how the body's relationship with objects nurtures science, a connection many resist. Psychoanalytic thinking addresses other questions about how objects inspire science. Here I consider how object *choice* and object *mastery* are determined by powerful inner forces, while object *space* creates a unique environment for creativity.

Object Choice

Walt Whitman captured something about how objects inspire when he said: "A child went forth every day/and the first object he look'd upon/that object he became."[2] But Whitman speaks about that first object as though it could be any object. In fact, the connections between child and object are specific and overdetermined.

In *Uncle Tungsten: A Memoir of a Chemical Boyhood,* Oliver Sacks describes the specificity of the object choices that led him to science. During World War II, Sacks, a Jewish child and a native of London, had been sent away with his brother Michael to boarding school in the country. His brother would leave the school broken, both physically and mentally; Oliver fared better, but only by degrees. When, at twelve, he returned to London, Sacks found objects that put him in contact with his worst fears and reassured him that they would not come to pass. A fearful object was aluminum smeared with mercury, which removed the aluminum's protective oxide coat. The surface of the aluminum erupted in "a white substance like a fungus . . . and it kept growing and growing until the aluminum was completely eaten up. . . . It made me think of a curse or a spell, the sort of disintegration I sometimes saw in my dreams.[3] For Sacks, the breakdown of the surface, the rotting away of the aluminum, evokes the death and carnage of the war, the cancer of fascism, anti-Semitism, the destruction of his brother's mind by encroaching madness.

While mercury was the metal of destruction, another metal represented safety and stability, promising that life, no matter what its limitations and restrictions, would, from that point forward, always stay the same. This metal was tungsten. Sacks had a beloved uncle, a chemist, who reassured him that tungsten can never be ravaged; mercury and its demons have no power over it. Sacks says: "If I put this little bar of tungsten in the mercury, it would not be affected at all. If I put it away for a million years, it would be just as bright and shiny as it is now."[4] Tungsten, at least, was stable in a precarious world.

The periodic table of the elements provides Sacks with another object of stability. Its order and symmetry are a balm to his spirit. The table gives him "for the first time, a sense of the transcendent power of the human mind, and the fact that it might be equipped to discover or decipher the deepest secrets of nature, to read the mind of God."[5]

Sacks's story is unique, but the overdetermination that characterize his object choices is not. The particular events of our lives inform the objects that compel us. So, for example, Timothy Bickmore, performing with the circus after his parents' divorce, puts on an attention-grabbing laser show to build a wall between himself and the audience. Lasers become his path into science. As a child, the molecular biologist Donald Ingber suffered from the conventionality of the suburb in which he was born. He puts Venus Paradise coloring pencils with their "color-by-number" constraints at the center of his rebellion. He defiantly colors outside of the lines. Through the pencils he meets the idea of the gestalt, where one final element makes a structure pop into three dimensions.

Play is children's work.[6] Children choose play objects that help them separate and individuate, that help them become their own people. Object work is deeply motivated; the emotion that fuels the investigations of young scientists taps into this intensity. This perspective brings us to a very different place than we would get to with a question such as: "What objects should children play with to learn science?" The object that brings a child to science doesn't have to be a Froebel gift or an electrical circuit. It has to be an object that speaks to a particular child. Not every object would have served Sacks as well as tungsten. Others could have been called into service, but tungsten had properties that made it unique—for Sacks.

Object Mastery

In object play, we have a chance both to discover and defy reality as it is presented to us. When this kind of mastery is turned toward the self, it is a step toward increased maturity.[7] When it is turned toward the world, we approach a scientific sensibility.

Children use object mastery to handle their earliest emotional challenges. We see this in Freud's description of a spool and string game invented by his one-and-a-half-year-old grandson. The child begins by making the spool disappear (calling it "*Fort,*" gone) and then bringing it back (now it is "*Da,*" there).[8] Freud theorized that this game of disappearance and return allowed the boy to manage his anxiety about the absences of his mother. By controlling the actual presence and absence of an object, he was able to represent his mother as a symbolic object, that is, represent his relationship with her. The boy came to terms with a concept—that his mother can be gone and yet still be present—that he symbolized in objects of play. In the play worlds closest to them and in the expanded worlds they share with others, children seek mastery of things outside themselves to put things right within themselves. The psychoanalyst Erik Erikson quotes William Blake on the importance of objects to childhood mastery: "The child's toys and the old man's reasons are the fruits of the two seasons."[9]

The child, Erikson continues, is trying to "deal with experience by creating model situations and to master reality by experiment and planning."[10] Such experiments can be done in the stillness of the laboratory or they can happen in the laughter of a backyard rumble. Janet Licini Connors and her friends used cardboard refrigerator boxes to perform a series of physics experiments. The experiments began with the children throwing their bodies around in the boxes. Then, the children improved the navigation of the boxes by taking off their flaps, turning them into tubes, and pushing them from the inside. Finally, the backyard physicists realized that things worked best when there was only one person inside of a box, crawling at a steady pace. Connors comments:

> We came to the conclusion that this crawling technique was the most efficient way to roll the box. It became our standard racing position. Looking back, I see that we were applying a scientific method. We tried different ways of moving the box, we made mistakes, and we looked at the results, which we measured in "box travel distance." This is my first memory of learning through physical action. Everything I know and understand I have learned this way.

Object Space

The boxes became an extension of the racers' bodies; otherwise put, the playing children learned to think of box space as body space. Like the spool and string of Freud's story, the boxes came to have a place in the children's inner and outer world. The psychoanalyst D. W. Winnicott calls such objects *transitional*.[11]

For Winnicott, the objects of the nursery (the stuffed animal, the favorite pillow) mediate between the child's sense of being part of the mother and being an independent self. These objects leave traces that will mark the rest of an individual's life. The joint allegiance of transitional objects to self and external world demonstrates to the child that objects in the external world can be loved. Winnicott believes that during all stages of life, we continue to search for objects we can love, objects we can experience as both within and outside ourselves. As adults, we divide the world into an inner and outer realm, and an "intermediate area of experiencing, to which inner reality and external life both contribute."[12] That intermediate area is creative and expressive. It is a space where science happens.

Scientists describe feeling both at one with and lost in nature when they are in this intermediate space. Objects help them reach it. Matthew Grenby, reflecting on soap bubbles and volcanoes, describes that space as one of "an almost imperceptible impression." Diane Willow, reflecting on soil and water, writes of "being able to think like the mud." Joanna Berzowska uses herself as "a solid geometer's object," her body and breath becoming ways to measure distances.

Gerald Sussman, a computer scientist at MIT, once described feeling so close to a pair of binoculars given to him when he was five that, when he thought of an idea, he thought that the binoculars shared in it. When he realized that he could look through his binoculars in both directions, Sussman came to a theory of reversibility, the idea that there were some processes that could work backward as well as forward.[13] He theorized that if you use gasoline to drive a car forward, it must be possible to get a car to generate gasoline if you run it backward, or put more formally, if you consume gasoline to generate motion, you should be able to consume motion to generate gasoline. Sussman gave the binoculars half credit for this idea and became confident that a partnership with objects would lead him to discoveries that, if not necessarily correct, would be thrilling.

The binocular-inspired theory of the reverse motor and Daniel Kornhauser's electric plant running on electrically charged shirts are stories from Winnicott's intermediate zone of relating, where the individual feels at one with larger forces. These children are not making efforts that lead to external reward or even external effect (no reverse motor ever produced any gasoline; no shirt sparks produced a boiler room to heat a cold French apartment). Rather, these stories provide a window onto how the minds of children develop into the minds of scientists, how object space becomes transitional space, where no idea is too "far out" and every idea can be made to feel part of the big picture.

Christopher Bollas, a psychoanalyst who works in Winnicott's tradition, analogizes transitional moments to aesthetic moments, breaks in experience during which "the subject feels held in symmetry and solitude by the spirit of the object."[14] And like Winnicott, Bollas makes it clear that once people have such object experiences, they search for them again and again, "The Christian may go to church and there hope to find traces of his experience, the naturalist may look for another sighting of that rarest of birds that creates for him a moment of sudden awe, and the romantic poet walk his landscape hoping for a spot in time, a suspended moment when self and object feel reciprocally enhancing and mutually informative."[15] Young scientists are inspired by objects in which they become lost on the way to finding themselves.

Objects do not determine the particular ideas they inspire. Sometimes the most important thing they inspire is the feeling of having a "charge," a "thrill" or a "secret theory" that leads children to want to have more. And five-year-olds with incorrect theories can be told there is more to learn, but they should be allowed to enjoy their creative charges.

When we read accounts of children discovering science through objects—Feynman's excitement and pride in his radios, Kornhauser's joy as he investigates his secret theories, Sacks's thrill when he meets tungsten and the periodic table—emotion infuses the space of discovery. One might even say that it *is* the space of discovery.

Finding Science

From mentors to students, the essays in this collection cover seventy-five years of experience with objects. In these narratives, objects have a tendency to slow things down. Scientific thinking that needed uninterrupted reflection flourished in worlds built around object passions. Stephen Intille's sand castles received the uninterrupted time of whole days and "a resource as boundless as a good stretch of beach and ocean." Selby Cull watches rocks that are "almost my children. I watch them, even when I don't have to."

Sand and minerals are natural objects; in this collection we see artifacts handled in the same patient spirit. Telephones are taken apart and put back together. Blocks worlds are built and rebuilt. Broken objects are repaired for the secrets they have to tell. Nights become long as a young mind contemplates how his shirt can make sparks. Susan Hockfield's essay on a love of microscopes captures how objects slow time:

Anything is interesting magnified. We kept the microscope in the kitchen so that we could look at something every day. We studied orange peels, leaves, flowers, bugs, newspaper print—anything and everything. You can easily spend an afternoon immersed in the intricacies of structure and how that structure leads to function.

These slow, reflective pleasures of the scientist are not so different from those of historians who inhabit other times and ways. What scientist and historian have in common is an experience that respects immersion rather than curricular pace. Their shared experience has little in common with lesson plans, accelerated drill and practice, or rapid-fire multiple simulations.

Digital media can be used to slow things down, to invite careful exploration, but in the virtual, velocity tempts because it is so easily achieved. Early personal computers invited their users to "open the hood" and look inside, but more recent digital media rarely seem to "want" to be used at leisure. Their great and unique virtue is that they are able to present an endless stream of what-ifs—thought experiments that try out possible branching structures of an argument or substitutions in an experimental procedure. At its heart, digital culture is about precision and an infinity of possibility. It is about creating a "second nature" under our control. When Andrew Sempere, a computer scientist, meets the primitive Holga camera in a high school art class, he describes it as having "all the mechanical accuracy and precision of a jar of peanut butter." The encounter is humbling. In a digital imaging class, Sempere complains about all the materials he doesn't have. The Holga inspires resourcefulness; it slows him down and he has to work with what is at hand. He comes to love what is at hand.

I believe that as such moments ground us, they open us, heart and mind, to fall for science. Children find physics in the collision of LEGO ships, mathematics in the motion of a fly rod, geology in the viscosity of a meringue. Objects inspire a passion for the particular. Children discover the stubborn complexity of soap bubbles and ask what kind of sand is best for building castles. In doing so, they may come to wonder at our Earth, not only as a frontier of science, but as where we live.

Notes

Sherry Turkle | *Introduction*

1. Seymour Papert and Alan Kay's essays were adapted, with their permission, from other writing; all other mentor essays were written for this volume. The student essays were originally written as class assignments and with permission have been edited for publication.
2. Thomas Friedman's *The World Is Flat* became a lightning rod for national debate on this issue (New York: Farrar, Straus and Giroux, 2005).
3. National Education Summit on High Schools. February 26, 2005. Prepared remarks by Bill Gates, chair. Available at <http://www.gatesfoundation.org/MediaCenter/Speeches/Co-ChairSpeeches/BillgSpeeches/BGSpeech NGA-050226.htm> (accessed January 18, 2008).
4. William J. Broad, "Top Advisory Panel Warns of an Erosion of the U.S. Competitive Edge in Science," *The New York Times,* sec. A, October 13, 2005, 22.
5. The idea that objects are working materials for thought is central to this volume, as it has been to my earlier work. It builds on Piagetian insights and Claude Lévi-Strauss's notion of *bricolage.* See Jean Piaget, *The Child's Construction of Reality,* trans. Margaret Cook (London: Routledge and Kegan Paul, 1955); *The Child's Conception of the World,* trans. John and Andrew Tomlinson (Totowa, N.J.: Littlefield Adams, 1960); Claude Lévi-Strauss, *The Savage Mind,* trans. John Weightman (Chicago: University of Chicago Press, 1966). See also Sherry Turkle, *The Second Self: Computers and the Human Spirit* (Cambridge, Mass.: MIT Press, 2005 [1984]) and Sherry Turkle, ed., *Evocative Objects: Things We Think With* (Cambridge, Mass.: MIT Press, 2007).

6. Seymour Papert, "The Gears of My Childhood," in *Mind-storms: Children, Computers, and Powerful Ideas* (New York: Basic Books, 1980), vi. In this collection, this preface is reprinted in the section on mentor essays.

7. Ibid., viii.

8. Cited in Michael Shortland and Richard Yeo, "Introduction" in *Telling Lives in Science,* Shortland and Yeo, eds. (Cambridge: Cambridge University Press, 1997), 8–9.

9. Ibid., 8. In this essay, the authors refer to Shortland's earlier work on scientific autobiography in which he cites the chemist Erwin Chargaff's opinion that scientific autobiography was likely to "lack personal interest" because contrary to the situation in the arts, "it is not the men that make science; it is the science that makes men." Michael Shortland, "Exemplary Lives: A Study of Scientific Autobiographies," *Science and Public Policy* 15 (1988): 172. In a recent book by Gregory J. Feist, *The Psychology of Science and the Origins of the Scientific Mind* (New Haven: Yale University Press, 2006), scientists' relationships with objects are not directly discussed at all.

10. Richard Feynman, *Surely You're Joking, Mr. Feynman!* [as told to Ralph Leighton; Edward Hutchins, ed.] (New York: W. W. Norton, 1985), 15.

11. Ibid., 8.

12. Papert, *Mindstorms.* See, especially, chapter 5, "Microworlds: Incubators for Knowledge," 120–134.

13. Nicholas Negroponte, *Being Digital* (New York: Knopf, 1995).

14. Lévi-Strauss, *The Savage Mind.*

15. Piaget, *The Child's Construction of Reality* and *The Child's Conception of the World.*

16. Idit Harel and Seymour Papert, eds., *Constructionism: Research Reports and Essays, 1985–1990* [by the Epistemology and Learning Research Group, The Media Laboratory, Massachusetts Institute of Technology] (Norwood, N.J.: Ablex, 1991).

17. See Sherry Turkle, *The Second Self.* See, especially, chapter 2, "Video Games and Computer Holding Power," 65–90.

18. Evelyn Fox Keller, *Gender and Science* (New Haven: Yale University Press, 1985).

19. Evelyn Fox Keller, *A Feeling for the Organism: the Life and Work of Barbara McClintock* (New York: W. H. Freeman, 1983), 198.

20. Ibid., 117.

21. D. W. Winnicott, *Playing and Reality* (London: Routledge, 1989).

22. I develop this developmental framework in greater depth in *The Second Self,* Part I, 33–152.

23. Sherry Turkle and Seymour Papert, "Epistemological Pluralism: Styles and Voices within Computer Culture," *Signs: Journal of Women in Culture and Society* 1, no. 16 (1990): 128–157. For a highly influential description of styles of mastery, see Howard Gardner, *Frames of Mind: The Theory of Multiple Intelligences* (New York: Basic Books, 1983).

24. Anthony is a pseudonym. My interview with him was part of a study that promised confidentiality.

25. Turkle, *The Second Self,* 216.

26. Ibid.

27. For a discussion of where objects fit in the Artificial Intelligence aesthetic of "building straight from the mind," see ibid., 230ff. Master game architect Will Wright (developer of the games in the Sim series: *SimAnt, SimCity, SimLife, The Sims, Sims Online*) is currently developing a game, now known as *Spore,* in which players develop a single-celled creature into a multicellular organism, design it to their specifications, and then put it in a computer-generated ecosystem, with its own nutrients, predators, opportunities, and threats. As the designed creature struggles to exist, it can evolve and move on to a next level where each player controls a group of creatures that form a primitive tribe, augmented by simple tools. Thus, from needing to be concerned about issues of basic metabolism, a player of *Spore* graduates to having to think about social dynamics (shall one's tribe be peaceful or warlike?), social priorities (shall one's tribe

exploit new technology or focus on low-tech social cama-
raderie?), and the issues raised by being part of a city, a
nation, a civilization, a planet, and, finally, an interga-
lactic universe. The game's aspiration to teach thinking
skills about complex process in a virtual environment is
emblematic of the "building straight from the mind" aesthetic. For a detailed description of the game, see Steven
Johnson, "The Long Zoom," *New York Times Magazine,*
October 8, 2006.

28. Turkle, *The Second Self,* especially chapter 3, "Child Pro-
grammers: The First Generation," and chapter 4, "Video
Games and Computer Holding Power"; and Turkle and
Papert, "Epistemological Pluralism."

29. See, for example, Piaget, *The Child's Construction of Reality* and *The Child's Conception of the World.*

30. Mitchel Resnick, <http://momentum.media.mit.edu/
resnick-long.html> (accessed August 18, 2007)

31. Mitchel Resnick, *Turtles, Termites, and Traffic Jams*
(Cambridge, Mass.: MIT Press, 1994). Adults, too, are re-
sistant to notions of what Resnick calls the "decentralized mindset." In observing demonstrations of Resnick's
work at the Boston Museum of Science, it is clear that
adults, no less than children, are for the first time meeting ideas about emergent phenomena in a form that they
can understand.

32. Ibid., 122.

33. Evelyn Fox Keller and Lee A. Segel, "Initiation of Slime
Mold Aggregation Viewed as an Instability," *Journal of
Theoretical Biology* 26 (1970).

34. StarLogo is an example of how a virtual laboratory can
diverge from this model; there, experiments are not performances; things "happen" rather than simply being
"found."

35. It encourages stereotypes that children are already developing on their own. In my investigations of gender and
computing—a crucial area, because in America, fewer
and fewer women are going into computer science each
year—I found a striking inversion of earlier patterns of

gender-based reticence about science. In the 1970s and 1980s, young women (from junior high school to college) said that they stayed out of computer and science courses because they were concerned about their own competence. Twenty-five years later, they were staying away from computers and science, but their reasoning had changed. It wasn't "We would like to get into science, but we don't think we can do it" but "We know we can do science, but we don't want to." They didn't want to because they associated scientific careers with isolation and not caring about people. Sherry Turkle, "Tech-Savvy: Educating Girls in the New Computer Age," coauthored with AAUW [American Association of University Women] Educational Foundation Commissioners, April 2000. Available at <http://www.aauw.org/research/TechSavvy .pdf> (accessed August 19, 2007).

36. Oliver Sacks, *Uncle Tungsten: A Chemical Boyhood* (New York: Vintage, 2000), 136.

37. Ibid.

Alan Kay | *The Vacuum*

1. This essay was adapted by the editor from Alan Kay's 2004 Kyoto Prize Commemorative Lecture, "The Center of Why," delivered on November 11, 2004. Available at <http://www.vpri.org/pdf/kprize_RN-2004-002.pdf> (accessed August 19. 2007).

Seymour Papert | *Gears*

1. This is an edited version of the essay published as the preface to Seymour Papert's *Mindstorms: Children, Computers, and Powerful Ideas* (New York: Basic Books, 1980).

Sherry Turkle | *Epilogue*

1. Lacan drew the circles in four colors, designating what he called the imaginary, the symbolic, and the real, and a fourth circle that he referred to as the *symptôme*. The cir-

cles were interlocking so that when one is cut, the whole chain of circles becomes undone.

For a fuller description of the Lacan visit, see Sherry Turkle, *Psychoanalytic Politics: Jacques Lacan and Freud's French Revolution* (Guilford, Conn.: Guilford Press, 1991 [1978]).

2. Walt Whitman, *Leaves of Grass* (New York: Random House, 1993 [1855]), 454.

3. Oliver Sacks, *Uncle Tungsten: Memories of a Chemical Boyhood* (New York: Knopf, 2001), 38–39.

4. Ibid., 39.

5. Ibid., 191.

6. Erik Erikson, *Childhood and Society* (New York: Norton, 1964), 222.

7. "The playing adult steps sideward into another reality; the playing child advances forward to new stages of mastery." Ibid.

8. Sigmund Freud, "Beyond the Pleasure Principle" [1920] in *The Standard Edition of the Complete Psychological Works of Sigmund Freud,* ed. and trans. James Strachey et al. (London: Hogarth, 1953–1974), vol. XVIII, 7–64.

9. Erikson, *Childhood and Society,* 222. Erikson also says, in relation to mastery:

> It is in certain phases of his work that the adult projects past experiences into dimensions which seem manageable. In the laboratory, on the stage, and on the drawing board, he relives the past and thus relieves leftover affects; in reconstructing the model situation, he redeems his failures and strengthens his hopes. He anticipates the future from the point of view of a corrected and shared past. (222)

10. Ibid.

11. D. W. Winnicott, "Transitional Objects and Transitional Phenomena: A Study of the First Not-Me Possession," *International Journal of Psychoanalysis* 34, no. 2: 89–97.

12. D. W. Winnicott, *Playing and Reality* (New York: Basic Books, 1971), 2.

13. Sussman continues:

> I was really shaken by that—that you could turn them around and it worked both ways. Somehow that clicked with other things—like if you look in a mirror and see someone, they can see you. I remember all of these things happened very fast. And all of a sudden I had this large collection of things and I suddenly realized that they were all the same thing.

> Quoted in Turkle, *The Second Self: Computers and the Human Spirit* (Cambridge, Mass.: MIT Press, 2005 [1984]), 231.

14. Christopher Bollas, *The Shadow of the Object: Psychoanalysis of the Unthought Known* (New York: Columbia University Press, 1987), 31.

15. Ibid.

Bibliography and Suggested Readings

From Biography, Memoir, and Scientists on Science

Bernstein, Jeremy. *Experiencing Science: Profiles in Discovery.* New York: E. P. Dutton, 1980.

————. *Science Observed: Essays Out of My Mind.* New York: Basic Books, 1982.

Brockman, John, ed., *Curious Minds: How a Child Becomes a Scientist.* New York: Pantheon Books 2004.

————, ed. *The Greatest Inventions of the Past 2000 Years.* New York: Simon and Schuster, 2000.

Browne, Janet. *Charles Darwin: Voyaging.* Princeton: Princeton University Press, 1995.

Darwin, Charles. *Autobiography.* London: Collins, 1958.

Davis, Philip J., and Reuben Hersh. *The Mathematical Experience.* Boston: Birkhauser, 1980.

Dyson, Freeman. *Disturbing the Universe.* New York: Basic Books, 1979.

Feynman, Richard. *Surely You're Joking, Mr. Feynman.* New York: W. W. Norton, 1981.

Flannery, Sarah, with David Flannery. *In Code: A Mathematical Journey.* Chapel Hill: Algonquin Books of Chapel Hill, 2003.

Gruber, Howard E. *Darwin on Man: A Psychological Study of Scientific Creativity.* New York: E. P. Dutton, 1974.

Hofstadter, Douglas. *Gödel, Escher, Bach: An Eternal Golden Braid.* New York: Basic Books, 1979.

Hillis, W. Daniel. *The Pattern on the Stone: The Simple Ideas That Make Computers Work*. New York: Basic Books, 1998.

John-Steiner, Vera. *Notebooks of the Mind: Explorations of Thinking*. Albuquerque: University of New Mexico, 1985.

Keller, Evelyn Fox. *A Feeling for the Organism: The Life and Work of Barbara McClintock*. San Francisco: W. H. Freeman, 1983.

Minsky, Marvin. *The Emotion Machine: Commonsense Thinking, Artificial Intelligence, and the Future of the Human Mind*. New York: Simon and Schuster, 2006.

Negroponte, Nicholas. *Being Digital*. New York: Vintage, 1996.

Roe, Anne. *The Making of a Scientist*. New York: Dodd, Mead and Co., 1952.

Rosenblith, Walter, ed. *Jerry Wiesner: Scientist, Statesman, Humanist: Memories and Memoirs*. Cambridge, Mass.: MIT Press, 2003.

Sacks, Oliver. *Uncle Tungsten: A Chemical Boyhood*. New York: Knopf, 2001.

Sayre, Anne. *Rosalind Franklin and DNA*. New York: W. W. Norton, 1975.

Shortland, Michael, and Richard Yeo, eds. *Telling Lives in Science: Essays on Scientific Biography*. Cambridge: Cambridge University Press, 1996.

Thomas, Lewis. *Lives of a Cell*. New York: Penguin Books, 1974.

Ulam, Stanislaw M. *Adventures of a Mathematician*. New York: Scribner, 1976.

Weisskopf, Victor F. *Knowledge and Wonder: The Natural World as Man Knows It.* Cambridge, Mass.: MIT Press, 1979.

Weizenbaum, J. *Computer Power and Human Reason: From Judgment to Calculation.* New York: W. H. Freeman, 1979.

Wiener, Norbert. M. *Ex-Prodigy: My Childhood and Youth.* Cambridge, Mass.: MIT Press, 1964.

Wright, Orville. *How We Invented the Airplane.* New York: David McKay, 1953.

From Philosophy and Psychology

Bakhtin, M. M. *The Dialogic Imagination: Four Essays.* Translated by Carl Emerson and Michael Holquist. Edited by Michael Holquist. Austin: University of Texas Press, 1981.

Belenky, Mary Field, Blythe Clinchy, Nancy Goldberger, and Jill Tarule. *Women's Ways of Knowing: The Development of Self, Voice, and Mind.* New York: Basic Books, 1986.

Bleier, Ruth, ed. *Feminist Approaches to Science.* New York: Pergamon, 1986.

Bergson, Henri. *Matter and Memory.* Translated by N. M. Paul and W. S. Palmer. New York: Zone Books, 1990.

Bower, T. G. R. *The Perceptual World of the Child.* Cambridge, Mass.: Harvard University Press, 1977.

Carey, Susan. *Conceptual Change in Childhood.* Cambridge, Mass.: MIT Press, 1987.

Feist, Gregory J. *The Psychology of Science and the Origins of the Scientific Mind.* New Haven: Yale University Press, 2006.

Gilligan, Carol. *In a Different Voice: Psychological Theory and Women's Development.* Cambridge, Mass.: Harvard University Press, 1982.

Gruber, Howard E. "On the Relation between 'Aha Experiences' and the Construction of Ideas." *History of Science* 19 (1981): 41–59.

Harding, Sandra, and Merrill B. Hintikka, eds. *Discovering Reality: Feminist Perspectives on Epistemology, Metaphysics, Methodology, and Philosophy of Science.* London: Reidel, 1983.

Heidegger, Martin. *The Question Concerning Technology, and Other Essays.* New York: Harper Torchbooks, 1977.

———. *What Is a Thing?* Chicago: H. Regnery Co., 1968.

Kagan, Jerome. *The Nature of the Child.* New York: Basic Books, 1984.

———. *Three Seductive Ideas.* Cambridge, Mass.: Harvard University Press, 1998.

Norman, Donald. *The Design of Everyday Things.* New York: Doubleday, 1990.

Pasztory, Esther. *Thinking with Things: Toward a New Vision of Art.* Austin: University of Texas, 2005.

Piaget, Jean. *The Child's Conception of the World.* Translated by Joan and Andrew Tomlinson. Totowa, N.J.: Littlefield Adams, 1960.

———. *Genetic Epistemology.* Translated by Eleanor Duckworth. New York: Columbia University Press, 1970.

———. *Intelligence and Affectivity: Their Relationship During Child Development.* Translated and edited by T. A. Brown and C. E. Kaegi. Palo Alto, Calif.: Annual Reviews, 1981.

Piaget, Jean, and Barbel Inhelder. *The Growth of Logical Thinking from Childhood to Adolescence.* New York: Basic Books, 1958.

Plato. *The Republic.* Translated by Allan Bloom. New York: Basic Books, 1968.

Sutton-Smith, Brian. *The Ambiguity of Play.* Cambridge, Mass.: Harvard University Press, 1997.

Vygotsky, Lev. *Mind in Society.* Cambridge, Mass.: Harvard University Press, 1978.

———. "Play and Its Role in the Mental Development of the Child." In *Play: Its Role in Development and Education,* edited by Jerome S. Bruner, Alison Jolly, and Kathy Sylva. New York: Penguin, 1976.

———. *Thought and Language,* edited by G. Vakar. Translated by E. Hanfmann. Cambridge, Mass.: MIT Press, 1962.

Wechsler, Judith, ed. *On Aesthetics in Science.* Cambridge, Mass.: MIT Press, 1977.

Weinberg, Gerald. *The Psychology of Computer Programming.* New York: Scribner, 1958.

From the Psychoanalytic Tradition

Bollas, Christopher. *The Shadow of the Object: Psychoanalysis of the Unthought Known.* New York: Columbia University Press, 1987.

Erikson, Erik H. *Childhood and Society.* New York: W. W. Norton, 1963.

Fonagy, Peter. *Attachment Theory and Psychoanalysis.* New York: Other Press, 2001.

Fraiberg, Selma H. *The Magic Years: Understanding and Handling the Problems of Early Childhood.* New York: Scribner, 1959.

Freud, Sigmund. *The Standard Edition of the Complete Works of Sigmund Freud,* edited by James Strachey, et al. London: The Hogarth Press and The Institute of Psychoanalysis, 1953–1974. See esp. "Fetishism, "The Interpretation of Dreams," "Mourning and Melancholia," "Parapraxes," "Totem and Taboo: Some Points of Agreement between the Mental Lives of Savages and Neurotics," and "The Uncanny."

Greenberg, Jay, and Stephen Mitchell. *Object Relations in Psychoanalytic Theory.* Cambridge, Mass.: Harvard University Press, 1983.

Klein, Melanie. *Love, Guilt and Reparation and Other Works, 1921–1945.* New York: Dell, 1975.

Lacan, Jacques. *Écrits: A Selection.* Translated by Bruce Fink. New York: W. W. Norton, 2002.

———. *The Four Fundamental Concepts of Psycho-Analysis.* Translated by Alan Sheridan. New York: W. W. Norton, 1978.

Phillips, Adam. *The Beast in the Nursery: On Curiosity and Other Appetites.* New York: Pantheon, 1998.

———. *On Kissing, Tickling, and Being Bored: Psychoanalytic Essays on the Unexamined Life.* Cambridge, Mass.: Harvard University Press, 1993.

Winnicott, D. W. *Collected Papers: Through Pediatrics to Psycho-Analysis.* London and New York: Tavistock Publications, Basic Books, 1958.

———. *Playing and Reality.* London: Routledge, 1989.

———. *Psycho-Analytic Explorations,* edited by Clare Winnicott, Ray Shepherd, and Madeleine Davis. Cambridge, Mass.: Harvard University Press, 1989.

———. "Transitional Objects and Transitional Phenomena: A Study of the First Not-Me Possession." *The*

International Journal of Psychoanalysis 34, 2:
89–97.

From the History and Social Studies of Science and Objects

Appadurai, Arjun, ed. *The Social Life of Things: Commodities in Cultural Perspective.* Cambridge: Cambridge University Press, 1988.

Briggs, Asa. "The Philosophy of the Eye: Spectacles, Cameras and the New Vision." *Victorian Things.* London: B. T. Batsford, 1988.

Csikszentmihalyi, Mihaly. *The Meaning of Things: Domestic Symbols and the Self.* Cambridge: Cambridge University Press, 1981.

———. "Why We Need Things." In *History and Things: Essays on Material Culture,* edited by Steven Lubar and W. David Kingery. Washington, D.C.: Smithsonian Institution Press, 1993.

Daston, Lorraine, ed. *Things That Talk: Object Lessons from Art and Science.* New York: Zone Books, 2004.

Daston, Lorraine, and Katharine Park. *Wonders and the Order of Nature: 1150–1750.* New York: Zone Books, 1998.

Galison, Peter. *Einstein's Clocks, Poincaré's Maps: Empires of Time.* New York: W. W. Norton, 2003.

———. *Image and Logic: A Material Culture of Microphysics.* Chicago: University of Chicago Press, 1997.

Jones, Caroline, and Peter Galison. *Picturing Science, Producing Art.* New York: Routledge, 1998.

Keller, Evelyn Fox. *Reflections on Gender and Science.* New Haven: Yale University Press, 1985.

Knorr Cetina, Karin. *Biographies of Scientific Objects.* Chicago: University of Chicago, 2000.

———. *Epistemic Cultures: How the Sciences Make Knowledge.* Cambridge, Mass.: Harvard University Press, 1999.

———. *The Manufacture of Knowledge: An Essay on the Constructivist and Contextual Nature of Science.* Oxford: Pergamon Press, 1981.

———. "Sociality with Objects: Social Relations in Post-social Knowledge Societies." *Theory, Culture, and Society* 14 (1997): 1–30.

Kuhn, Thomas S. *The Structure of Scientific Revolutions.* 2d ed. Chicago: University of Chicago Press, 1970.

Latour, Bruno. *Science in Action: How to Follow Scientists and Engineers through Society.* Cambridge, Mass.: Harvard University Press, 1987.

———. *We Have Never Been Modern.* Translated by Catherine Porter. Cambridge, Mass.: Harvard University Press, 1993.

Latour, Bruno, and Steven Woolgar. *Laboratory Life: The Social Construction of Scientific Facts.* Princeton, N.J.: Princeton University Press, 1979.

Lynch, Michael, and Steven Woolgar, eds. *Representation in Scientific Practice.* Cambridge, Mass.: MIT Press, 1990.

Mumford, Lewis. "The Monastery and the Clock." In *Technics and Civilization.* New York: Harcourt, Brace and Company, Harvest, 1963.

Pickering, Andrew. *The Mangle of Practice: Time, Agency, and Science.* Chicago: University of Chicago Press, 1997.

Pinch, Trevor, and Frank Trocco. *Analog Days: The Invention and Impact of the Moog Synthesizer.* Cambridge, Mass.: Harvard University Press, 2002.

Star, Susan Leigh, and Geoff Bowker. *Sorting Things Out: Classification and Its Consequences.* Cambridge, Mass.: MIT Press, 2000.

Strohecker, Carol. Why Knot? PhD diss. MIT, Cambridge, Mass., 1991.

Suchman, Lucy. "Affiliative Objects." *Organization* 12, no. 3 (2005): 379–399.

———. *Human-Machine Reconfigurations: Plans and Situated Actions.* 2nd expanded ed. New York: Cambridge University Press, 2007.

Turkle, Sherry. *Life on the Screen: Identity in the Age of the Internet.* New York: Simon and Schuster, 1995.

———. *The Second Self: Computers and the Human Spirit.* 20th anniversary ed. Cambridge, Mass.: MIT Press, 2005.

———. "Whither Psychoanalysis in Computer Culture." *Psychoanalytic Psychology: Journal of the Division of Psychoanalysis* 1, no. 21 (2004): 16–30.

Traweek, Sharon. *Beamtimes and Lifetimes: The World of High Energy Physicists.* Cambridge, Mass.: Harvard University Press, 1988.

From Classical Social Theory

Berger, Peter L., and Thomas Luckmann. *The Social Construction of Reality: A Treatise in the Sociology of Knowledge.* Garden City, N.Y.: Doubleday, 1966.

Douglas, Mary. *Purity and Danger: An Analysis of Concepts of Pollution and Taboo.* New York: Routledge, 1966.

Durkheim, Emile. *The Elementary Forms of the Religious Life.* Translated by J. W. Swain. London: G. Allen and Unwin, 1995.

Durkheim, Emile, and Marcel Mauss. *Primitive Classification.* Translated by Rodney Needham. Chicago: University of Chicago Press, 1963.

Geertz, Clifford. *The Interpretation of Cultures: Selected Essays.* New York: Basic Books, 1973.

Levi-Strauss, Claude. *The Savage Mind.* Translated by John Weightman and Doreen Weightman. Chicago: University of Chicago Press, 1966.

―――. *Structural Anthropology.* Translated by Monique Layton. New York: Basic Books, 1963.

―――. *Totemism.* Translated by Rodney Needham. Boston: Beacon Press, 1963.

―――. *Tristes Tropiques.* Translated by Doreen Weightman and John Weightman. New York: Atheneum, 1974.

Mauss, Marcel. *The Gift: The Form and Reason for Exchange in Archaic Societies.* Translated by W. D. Halls. New York: W. W. Norton, 2000.

Turner, Victor Witter. *The Forest of Symbols: Aspects of Ndembu Ritual.* Ithaca, N.Y.: Cornell University Press, 1967.

―――. *The Ritual Process: Structure and Anti-Structure.* Chicago: Aldine, 1969.

van Gennep, Arnold. *Rites of Passage.* Translated by Monika B. Vizedome and Gabrielle L. Caffee. Chicago: University of Chicago Press, 1960.

From Literary and Cultural Studies

Abraham, R. M. *Easy-to-Do Entertainments and Diversions with Coins, Cards, String, Paper, and Matches.* New York: Dover, 1961.

Aries, Philippe. *Centuries of Childhood: A Social History of Family Life.* Translated by Robert Baldick. New York: Knopf, 1962.

Amato, Ivan. *Stuff: The Materials the World Is Made Of.* New York: Basic Books, 1997.

Barthes, Roland. *Mythologies.* Translated by Annette Lavers. New York: Hill and Wang, 1972.

Baudrillard, Jean. *Simulacra and Simulations.* Translated by Sheila Faria Glaser. Ann Arbor: University of Michigan Press, 1995.

———. *The System of Objects.* London: Verso, 1996.

Benjamin, Walter. *Illuminations.* Translated by H. Zohn. Edited by Hannah Arendt. New York: Harcourt, 1968.

Brant, Sandra, and Elissa Cullman. *Small Folk: A Celebration of Childhood in America.* New York: E. P. Dutton, 1980.

Brown, Bill, ed. *Things.* Chicago: University of Chicago Press, 2004.

Fraser, Antonia. *A History of Toys.* New York: Delacorte, 1966.

Grober, Karl. *Children's Toys of Bygone Days.* New York: Frederick A. Stokes Company, 1928.

Hayles, Katherine N. "Simulated Nature and Natural Simulations: Rethinking the Relations between the Beholder and the World." In *Uncommon Ground,* edited by W. Cronin. New York: W. W. Norton, 1995.

Jenkins, Henry, and Justine Cassell, eds. *The Children's Culture Reader.* New York: New York University Press, 1998.

King, Constance Eileen. *The Encyclopedia of Toys.* New York: Crown, 1978.

Koestler, Arthur. *The Act of Creation.* New York: Dell, 1964.

McLuhan, Marshall. *Understanding Media: The Extensions of Man.* New York: American Library, 1964.

Nelms, Henning. *Thinking with a Pencil.* New York: Barnes and Noble, 1969.

Petroski, Henry. *The Evolution of Useful Things.* New York: Knopf, 1993.

———. *The Pencil: A History of Design and Circumstance.* New York: Knopf, 1989.

Pirsig, Robert. *Zen and the Art of Motorcycle Maintenance: An Inquiry into Values.* New York: Bantam Books, 1975.

Stafford, Barbara Maria, and Frances Terpak. *Devices of Wonder: From the World in a Box to Images on a Screen.* Los Angeles: Getty Publications, 2001.

Tenner, Edward. *Our Own Devices: The Past and Future of Body Technology.* New York: Knopf, 2003.

———. *When Things Bite Back.* New York: Knopf, 1996.

Ullman, Ellen. *Close to the Machine: Technophilia and Its Discontents.* San Francisco: City Lights Books, 1997.

From Writings on Education

Almquist, Birgitta. "Educational Toys, Creative Toys." In *Toys, Play and Child Development,* edited by J. H. Goldstein. New York: Cambridge University Press, 1994.

Brosterman, Norman. *Inventing Kindergarten.* New York: Harry N. Abrams, 1997.

Bruner, Jerome S. *The Process of Education.* New York: Vintage, 1963.

Dewey, John. *Democracy and Education: An Introduction to the Philosophy of Education.* New York: The Free Press, 1968.

DiSessa, Andrea A. *Changing Minds: Computers, Learning, and Literacy.* Cambridge, Mass.: MIT Press, 2001.

Froebel, Friedrich. *The Education of Man.* Translated by W. N. Hailmann. Mineola, N.Y.: Dover, 2005.

Gardner, Howard. *Five Minds for the Future.* Cambridge, Mass.: Harvard Business School Press, 2006.

————. *Frames of Mind: The Theory of Multiple Intelligences.* New York: Basic Books, 1983.

Goldman-Segall, Ricki. *Points of Viewing Children's Thinking: A Digital Ethnographic Journey.* Mahwah, N.J.: Lawrence Erlbaum, 1998.

Harel, Idit, and Seymour Papert, eds. *Constructionism: Research Reports and Essays, 1985–90, by the* MIT Epistemology and Learning Group, the MIT Media Laboratory. Norwood, N.J.: Ablex, 1991.

Holt, John. *How Children Learn.* New York: Dell, 1983.

Kay, Alan. "Powerful Ideas Need Love Too!" Joint Hearing on Educational Technology in the 21st Century, Science Committee and the Economic and Educational Opportunities Committee, US House of Representatives, October 12, 1995. Washington, D.C.

Lave, Jean. *Cognition in Practice: Mind, Mathematics and Culture in Everyday Life.* Cambridge: Cambridge University Press, 1988.

Lillard, Paula Polk. *Montessori: A Modern Approach.* New York: Schocken Books, 1972.

Papert, Seymour. *The Children's Machine: Rethinking School in the Age of the Computer.* New York: Basic Books, 1993.

———. *Mindstorms.* New York: Basic Books, 1980.

Resnick, Mitchel. "Technologies for Lifelong Kindergarten." *Educational Technology Research and Development* 46, no. 4 (1998): 43–55.

Resnick, Mitchel. *Turtles, Termites, and Traffic Jams: Explorations in Massively Parallel Microworlds.* Cambridge, Mass.: MIT Press, 1994.

Rousseau, Jean Jacques. *Emile: Selections.* Translated and edited by William Boyd. New York: Columbia University Press, 1969 [1762].

Schwebel, Milton, and Jane Raph, eds. *Piaget in the Classroom.* New York: Basic Books, 1973.

Starr, Paul. "Seductions of Sims: Policy as a Simulation Game." *The American Prospect* 17 (Spring 1994): 19–29.

Turkle, Sherry, and Seymour Papert. "Epistemological Pluralism: Styles and Voices within the Computer Culture." *Signs: Journal of Women in Culture and Society* 1, no. 16 (1990): 128–157.

Weir, Sylvia. *Cultivating Minds: A Logo Casebook.* New York: Harper and Row, 1987.

Illustration Credits

Exploding pods photo on pages 2 and 272 courtesy of Gilbert Raff; photo by Gilbert Raff.

Holga camera photo on page 41 (top) from Wikipedia.com. Photo by Mark Wheeler. Licensed under the Creative Commons Attribution ShareAlike 2.5 License.

Holga photo on page 41 (bottom, left) courtesy of Wikipedia.com. Sample photograph from a Holga camera showing vignetting and color saturation. Created on Nov. 11, 2005 by Clngre. This file is licensed under the Creative Commons Attribution ShareAlike license versions 2.5, 2.0, and 1.0.

Holga photo on page 41 (bottom, right) from Wikipedia .com. A photo by killerbear (May '06). Licensed under Creative Commons Attribution 2.5 License.

LEGOs photo on page 131 by Erin Hasley and Alyssa Larose. LEGOs courtesy of Anna Prior.

Wallpaper photo on page 167 by Erin Hasley and Alyssa Larose.

Apple II photo on page 179 courtesy of Steven Stengel, http://oldcomputers.net.

Radio photo on page 247 courtesy of Don Norman.

Pneumatic tube photo on page 261 from http://www .ids.u-net.com/cash/pneu-terminals.htm. Photo by Rod Moore.

Note: All illustrations without specific credit are copyright-free stock photos, are public domain, or were created by the book's art director.

Index

Aesthetics, objects and, 37, 38, 49–51, 68, 103, 106, 124, 137–141, 145–146, 156–157, 180, 222–223, 230, 236–240, 246, 254–259, 263

Aguirre, Lauren Seeley, 70–72

Alibaruho, Kwatsi, 14, 15, 18, 23, 134–136

Aluminum, 275–276

Alvarado, Christine, 3, 80–81

Analog to digital, 26–36, 49–51, 282

"Anthony," 27–28

Apple II, 11, 183–188, 206–207

Assembly (and machine) language, 11, 181–182, 196, 200

Assimilation (Piaget), 269

Atari 800, 203–205

Atari 2600, 10–11, 189–192

Bagehot, Walter, 8

Barbie doll, 59

Baking, 73–75, 96–99

BASIC (programming language), 193–196, 199–202, 203, 206

BASIC manual, 180–182

Batteries, 34, 35, 147, 212, 213, 214, 262

BBS (bulletin board system). *See* Modem

Beaudin, Jennifer, 12, 47–48.

Beehives, 7, 237, 239

Benton, Jonah, 105–107

Berzowska, Joanna, 58–59, 279

Bickmore, Timothy, 11, 142–146, 276

Bikes (bicycles), 14, 18, 23, 132–133, 135–136, 255

Binoculars, 279–280

Blake, William, 278

Blocks, 10, 16, 17, 18, 19, 35, 153, 180, 242–246, 281. *See also* "What We Build"

Body involvement, 20, 21, 29, 32, 47–48, 58–59, 114–117, 190–191, 232, 270. *See also* Closeness to the object; Object, intimacy; Tinkering (*bricolage*)

Bollas, Christopher, 280, 290nn14, 15

Bonk's Adventure (game), 214

Boxes, cardboard, 114–117, 278–279

"Box racing," 115–116

Brave, Scott, 159–161

Bricolage. See Closeness to the object; Styles of mastery, tinkering vs. planning; Tinkering (*bricolage*)

Brio, 164

Broad, William J., 284n4

Bubbles, 33–34, 60–62, 279, 283

"Building straight from the mind," 27, 286n27

Burglar alarms, 9

Calzaretta, Joseph, 173–174

Cameras, 229. *See also* Holga

Canadian Rockies, 223

Cancer research, 258

Cards, 74, 175–177

Carmel, Erica, 108–110

Carpenter's ruler, 20

Cars, 25, 134, 154, 159–161, 186, 268

Centipede (game), 206

Centripetal force, 108–110

Chapman, Robbin, 123–125

Chargaff, Erwin, 285n9

Chocolate meringue, 10, 11, 96–99

Choi, Eric, 111–113

Chu, Andrew, 156–158

Circus, 142–145

Clay, 16, 60, 63–65, 239

Clocks, 27, 76–79, 147

Closeness to the object, 18–22, 46, 68, 136, 181–182, 191, 200, 222, 238. *See also* Body involvement; Object, intimacy; Objects, personal and social identity and; Tinkering (*bricolage*)

Collection, pleasures of, 9, 42, 44, 111–113, 120–122, 147–166, 242–246

Computers, 5, 9–11, 27–32, 36, 94, 113, 119, 149, 166, 180–215, 246, 250, 271, 273, 282

Connors, Janet Licini, 114–117, 278–279

Conservation of energy, 33

Constructionism, 15–16

Constructivism, 15–16

Competition, 114–117, 123–125, 190

Cooking, 73–75

Cords, 102–104

Cull, Selby, 10, 11–12, 34, 96–99, 281

Dams, 16, 61, 239

DC ModemPak. *See* Modem

De Bonte, Austina (Vainius), 21, 23–24, 32, 137–141

Debugging, 11, 23, 140, 141, 181, 205, 207

Dice, 105–107

DNA, 256

Dodge, Chris, 193–198

Dollhouses, 15, 48, 221

Dolls, 3, 59, 70–72, 80–81, 108

Drawing vs. modeling, 16, 239

Dubner, Fran, 231

Dungeons and Dragons (game), 183

Dyslexia, 63

Eames, Charles, 246

Easy-Bake Oven, 73–75

Egg basket, 108–110

Electromagnetism, 45–46

Eltringham, Sandie, 25, 150–152

Emergent phenomenon, 30–31, 201–202

Entropy, 33, 90, 174

Erector Set, 54, 134–136, 255, 268

Erikson, Erik, 278, 289nn6, 7, 9, 290n10

Esserman, Chuck, 23, 132–133

Experiment, 11, 30, 31, 34, 64, 72, 86, 88–90, 96, 97, 103, 108–110, 115, 134, 138, 140, 144, 149, 157, 160, 220, 228, 230, 242, 278, 282

False starts (frustrations), learning from, 33, 51, 56 57, 59, 60, 82, 88, 90, 112, 135, 140, 151, 160, 177, 195, 214, 230, 249–250, 273. *See also* Debugging

Fan, 15, 85–86, 264–265

Fantasy football (game), 105–107

Feedback, 30, 82

Feist, Gregory J., 285n9

Fernbank Science Center, 231

Feynman, Richard, 9–10, 11, 15, 16, 281, 285nn10, 11

FidoNet, 211

"Fitness to purpose," 237

Fly rod (line), 12, 82–83

For Everyone a Garden (Safdie), 239

"Fort-Da" (game), 277–278

Freud, Sigmund, 277–278, 289n8

Friedman, Thomas, 284n2

Froebel, Friedrich, 16–17, 21, 31, 242, 277

Frogger (game), 206

Fuller, Buckminster, 157

Function (in mathematics), 82–83, 205

Functional abstraction, 201–202

Fuses, 9

Gardens, 238–239

Gardner, Howard, 286n23

Garvey, Steve, 106

Gates, Bill, 4

Gears, 3, 5, 6, 10, 22, 34, 133, 154, 262, 268–271, 273–274. *See also* Papert, Seymour

Gender, 20, 37, 72, 74, 80, 123, 233

Geodesic dome, 257

Georgia Tech, 233

Gestalt, 12, 256

GI Joe, 162, 194

Gravity, 27, 58, 110, 157, 258
  center of, 157–158

Grenby, Matthew, 33–34, 60–62, 279

Guitar, 136, 162

Gumby, 70–72

Hadar Hacarmel, 236
Hauer, Erwin, 257–258
Heat
  diffusion of, 84–87
  generation of, 88–91
Hermitt, Thomas P., 20, 45–46
Hockfield, Susan, 13, 15, 19, 20, 220–226, 281–282
Holga, 10, 49–51, 282. *See also* Camera
Hologram, 232

Ingber, Donald, 12, 252–259, 276–277
Internet, 207, 211
Intille, Stephen, 54–57, 281

Jacks, 123–125
Johnson, Steven, 287n27

Kaboom! (game), 190
Kahn, Louis, 237
Kay, Alan, 34–36, 262–265, 284n1, 288n1
Kaye, Joseph "Jofish," 164–166
Keller, Evelyn Fox, 18–19, 20, 31, 286nn18, 19, 20, 287n33
Keys, 111–113
Kiang, Douglas, 126–129
Kindergarten, 16, 30, 242, 244
Knots, 274–275
Knowledge, fragility of, 92–95
Kornhauser, Daniel, 32–33, 88–91, 280, 281
Kuhn, Sarah, 16, 17, 18, 19, 23, 242–246

Lacan, Jacques, 274–275, 289n1
Lamp bank, 9, 11
Language, systems of, 42–44
Lasers, 7, 10, 11, 23, 142–146, 232, 233, 276
Lebwohl, Rachel Elkin, 206–207
LEDS (light-emitting diodes), 212
Lee, Kwan Hong, 120–122

LEGOs, 7–8, 14, 24–25, 30, 35, 54, 55, 134–136, 147–166, 190
  categories, 152, 164–166
  laws, 156–158
  metrics, 150–152
  particles, 147–149
  people, 149, 151, 153–155, 164
  planning, 152, 159–161
  replicas,162–163
Lemonade Stand (game), 199
Lévi-Strauss, Claude, 14, 15, 284n14
Lewin, Walter, 109–110
"Lifelong Kindergarten," 30, 31
Light bulbs, 9, 11
Lincoln Logs, 48, 180, 255
Liu, Alan, 25, 153–155
Locks, 111–113
Logo (programming language), 28–30, 166, 206

Machal-Cajigas, Antoinne, 212–215
Magnifying glass, 15, 88, 220
Mapping the Atari, 205
Maps (mapping), 12, 22, 35, 42–44, 47–48, 61, 159
Marble, Justin, 147–149
Marbles, 5, 20, 120–122
Marcovitch, Emmanuel, 76–79
Marlow, Cameron, 12, 82–83
Martin, Fred, 180–182
Mathland, 28
McClintock, Barbara, 19, 20
Memoir, role in science and science education, 38, 273–274
Mentor essays, 6, 7, 218–271
Meteorite, 98
Microscope, 13, 19–20, 220–226, 281–282
Microworlds, 9, 12, 15–18, 27, 28, 36, 38, 42–44, 60, 105–107, 132–133, 135, 137–141, 148, 153–155, 182, 199, 236–240, 242–246, 249, 252–254, 264, 273. *See also* Mathland; Objects-to-think-with; Papert, Seymour; "What We Build"; "What We Sort"
  LEGOs as, 147–166 (*see also* LEGOs)

Minar, Nelson, 199–202

Mindstorms: Children, Computers, and Powerful Ideas
    (Papert), 273

Missile Command (game), 189–190, 192

MIT, 3–5, 6, 14, 29, 72, 225, 274
  class assignment on objects, 6, 10
  Media Lab, 64

Model, as part of design process, 239–240

Modem, 187, 208–211

Mondrian, Piet, 254

Montessori, 271

Motors, 147, 264, 280

Mud, 66–68, 279

Murtaugh, Michael, 73–75

Museum of Modern Art, 254

Music box, 118–119

My Little Pony, 3, 82–83

National Academies, 4

National Institutes of Health, 223, 225

Nature, 13, 21, 33–34, 45–46, 54–57, 60–62, 67, 97, 221,
    222, 223, 256–258, 283. *See also* "What We Sense"

Negroponte, Nicholas, 13

Nesheim, Britt, 22, 170–172

Newton, Sir Isaac, 45

Niemczyk, Steve, 203–205

Norman, Donald, 13–14, 15, 248–250

Novash, Walter, 102–104

Object
  assembly and disassembly, 13–15, 16, 18, 21–22, 134–136,
    147–149, 150–152, 156–158, 248–249, 255 *(see also*
    Thinking, analytic and procedural, objects on path to;
    Tinkering [*bricolage*])
  choice, 6, 24–25, 275–277
  intimacy, 12, 18–22, 37, 182, 185, 191, 243 *(see also* Body
    involvement; Closeness to the object; Objects, personal
    and social identity and; Tinkering [*bricolage*])
  play, 8, 30, 61, 102–104, 105–107, 154, 170–171, 175–177,
    184–185, 190, 245–246, 277 *(see also* Closeness to the

object; Microworlds, Object, intimacy; "Objects in Mind";
    Objects-to-think-with, Tinkering [*bricolage*])
  space, 35, 55, 59, 279–281 (*see also* Transitional object)
 Objects
  broken, 3, 9, 10, 34, 49, 50, 118–119, 192, 255, 281
  compelling nature of ("object passion"), 3, 5–6, 8, 9, 13,
    25, 35, 36, 37, 38, 45, 49–51, 106, 145, 173, 185–186,
    189–190, 205, 206, 212–213, 224, 231–233, 249, 270–
    271, 273–274, 279 (*see also* Closeness to the object;
    Microworlds; Object, intimacy; Transitional object)
  digital and virtual, 8, 10, 11, 12, 13, 18, 26–36, 76, 93, 211,
    282 (*see also* "What We Program")
  obsolete, 26–27, 92–95
  personal and social identity and, 6, 9–10, 16, 21, 23, 32, 37,
    43–44, 63, 64, 73–74, 112, 123, 132–133, 136, 145, 150,
    171, 183, 186, 188, 191, 194, 206, 212, 233, 239–240, 249
 "Objects in Mind," 268–271
 Objects-to-think-with, 5, 9, 15, 20, 28, 76–78, 80–81, 83,
    84–87, 88–91, 92–95, 97–98, 102–104, 109, 120–122, 128,
    161, 194, 197, 201–202, 205, 211, 213–214, 224–225,
    237, 242, 250, 252–258, 264–265, 268–271. *See also*
    Microworlds; Tinkering (*bricolage*)
 On Growth and Form (Thompson), 237–238
 Opacity, 11, 26, 71, 92–95, 197–198
 Ownership, sense of, 33, 56, 57, 74, 90,109, 113, 143–145,
    182, 215, 221, 228, 232, 240, 243, 250, 255, 262, 269.
    See also Secrets

 Pachinko machine, 126–129
 Pac-Man (game), 191
 Papert, Seymour, 5, 6, 9, 10,15, 21, 22, 28–29, 30, 268–271,
    273–274, 284n1, 286n23, 287n28, 288n1
 Patten, James, 84–87
 PCB (printed circuit board), 212–215
 Periodic table, 38, 276
 Peretti, Jonah, 63–65
 Photographs, 37–38, 49–51
 Piaget, Jean, 15, 16, 28, 269–270, 274, 284n15, 287n29
 Picard, Rosalind, 23, 228–233

Planning, 18, 43, 56, 64, 139–141, 159–161

Playmobil, 164

Primitivity, virtues of, 49–51

Prisms, 20, 45–46, 283

Programming, computer, 11, 27–31, 57, 74, 136, 149, 163, 180–182, 193–198, 199–202, 203–205, 206–207

Projector, 142–145

Psychoanalytic ideas, 20, 273–282. *See also* Bollas, Christopher; Erikson, Erik; Freud, Sigmund; Lacan, Jacques; Transitional object; Winnicott, D. W.

Purple haze chemistry, 228–233

Quirk, Miss Mary, 34, 35, 36, 262–264

Radio, 3, 5, 7, 9–10, 13, 16, 113, 147–148, 162, 248–250

Radio Amateurs, 249

Radio Shack, 180–181, 208

"Real, resistance of the," 11, 34, 49–51, 56, 60, 61, 282

Record player, 26, 92–93, 147

Recursion, 3, 80–81

Relationships

  family, 6, 12, 16, 26, 33, 35–36, 37, 43–44, 49, 54–55, 60, 67, 70, 72, 73, 78, 81, 82, 84, 85, 88, 90, 95, 96, 109, 111, 114, 126, 134–135, 143, 145, 150–152, 159, 164, 168, 173–174, 175–177, 183, 185–187, 189, 191, 193, 194, 195, 203–205, 206–207, 212–215, 221, 224, 228, 229, 237–240, 243–246, 252, 254, 255, 265, 269, 276, 277–278, 279

  friends, 16, 35, 48, 55, 114–117, 120, 123–125, 135, 145, 156, 164, 183–186, 190, 199, 220, 229, 231–232, 239, 278

  teachers, 34–35, 36, 37, 83, 109–110, 207, 228, 229, 230, 231–232, 233, 255, 262–264

Resnick, Mitchel, 29–31, 287nn30, 31

Robots, 134, 135, 156

Rocks, 281

Rogers, Buck, 134

Rorschach test, 254

Sacks, Oliver, 37–38, 275–276, 281, 288nn36, 37, 289nn3, 4, 5

Safdie, Moshe, 10, 16, 236–240

Safety and/or control, feeling of, 17, 22–23, 37–38, 43, 63–65, 74, 105, 145, 152, 183, 186–187, 205, 244–245, 255, 276

Sand castles, 5, 54–57, 281, 283

School, 34–36, 55, 64, 83, 97, 149, 157, 180, 185, 199, 204, 228–233, 238, 239, 256–257, 262–264, 268. *See also* Relationships, teachers

Schwartz, Steven, 22, 42–44

Science, cookbook style of, 228, 230, 233

Science Committee of the House of Representatives, 4

Science education, 4, 8, 18, 28–38, 49–51, 94–95, 256–259, 262–265, 268–271, 273–283

Science and technology, magic and mystery of, 14, 34, 74, 89, 90, 109, 142, 183, 196, 200, 209, 248–250. *See also* Secrets

Scientific disciplines, objects on the path to

anatomy, 220–226

biology, 30, 201, 220–226, 252–259

chemistry, 23, 38, 201, 230, 275–276

computer science, 57, 74, 107, 113, 119, 125, 136, 145, 149, 163, 180–215, 228–233, 246, 268–271, 279–280

design, 16–17, 23, 25, 44, 236–240, 242–246, 262–265

engineering, 7, 13, 26, 57, 94, 95, 139, 141,155, 248–250, 259

geology, 11–12, 97–99

mathematics (including geometry and statistics), 3, 5, 12, 28–30, 59, 76–78, 80–81, 83,103, 107, 116, 120–121, 125, 128, 137–141, 161, 162, 163, 172, 174, 205, 249, 263, 268–271, 274–275

physics, 5, 20, 45–46, 58, 63–64, 83, 84–87, 90, 103, 105–107, 116, 121, 128, 157–158, 174, 201, 232–233, 249, 265, 278

psychology, 13, 30, 248–250

Scratch (computer language), 31

Sculpture, 256–257

Secrets, 90, 143–144, 192, 281. *See also* Ownership, sense of

Segel, Lee A., 287n33

Sempere, Andrew, 10, 49–51, 282

Ship design, 156–158

Shirts, 33, 88–91, 280

Shortland, Michael, 285nn8, 9

Siaudinukai, 21, 23, 32, 137–141

SimAnt (game), 286n27

SimCity (game), 286n27

SimLife (game), 286n27

Simplicity and complexity, 147–149

Sims, The (game), 286n27

Sims Online (game), 286n27

Skydiving, 232–233

Snelson, Kenneth, 257

Sorting, 148, 151–152, 165. *See also* "What We Sort"

Space Invaders (game), 195–196

SpectraPhysics, 144

Spiegel, Dana, 162–163

Spore (game), 286n27

Sputnik, 138

StarLogo (programming language), 31, 287n34

Star Raiders (game), 203

Star Trek, 134

Star Wars, 157

Steps (terraces), 7, 16, 236–240

Stop signs, 173–174

Story, David (Duis), 183–188

Strauss, Todd, 168–169

Straws, 21, 22, 23, 24, 32, 137–141

Styles of mastery, tinkering vs. planning, 18. *See also*
 Closeness to the object; Planning; Thinking, analytic and
 procedural, object on the path to; Tinkering (*bricolage*);
 Surely You're Joking, Mr Feynman! (Feynman), 9

Sussman, Gerald, 279–280, 290n13

Symbols, objects as path toward, 173–174, 175–177, 194

Symmetry, 21, 58–59, 139–141

Technion, 236

Telephones, 14, 19, 35, 135, 162

Television sets, 147, 148

Tensegrity, 257–258

Texas Instruments 99/4A, 208

Thinking, analytic and procedural, objects on path to, 7–8, 12, 18, 20, 22–25, 31, 38, 64, 74, 105–107, 109, 115–117, 124–125, 136, 147–149, 150–152, 157–158, 159–161, 162–163, 172, 174, 180–182, 194, 196, 201, 204–205, 214, 221, 230, 233, 264, 268–271, 279–281. *See also* Experiment

Thompson, D'Arcy, 237–238

Tiller, Dr., 231–232

Tinkering (*bricolage*), 14, 18, 24, 54, 71–72, 138–141, 195–197, 201. *See also* Closeness to the object; Object, intimacy

Tinkertoys, 54, 55, 255, 288n35

Tivol, Brian, 175–177

Townes, Dr. Charles H., 233

Townsend, Anthony, 208–211

Toy mailbox, 22, 170–172

Transitional object, 20, 270, 279–280. *See also* Transitional space; Winnicott, D. W.

Transitional space, 34, 61–62, 168–169, 184, 191. *See also* Bollas, Christopher; Transitional object; Winnicott, D. W.

Transparency, 10–15, 26, 27, 71, 93, 113, 135–136, 142–146, 147, 186, 191–192, 197–198, 199–201, 204–205, 213–214, 220–221, 248–250

TRS-80, 10, 193–198

Tungsten, 276

TurboGrafx 16, 212–215

Turkle, Sherry, 3–38, 273–283, 284n5, 285n17, 286nn23, 25, 287n28, 288n1, 290n13

Turtle (Logo), 28–29, 30

Twirling, 102–104

Typewriter, 147, 244

Ultima (game), 199

Uncle Tungsten (Sacks), 37–38, 275–276

Vacuum (physical), 46, 264–265

Vacuums, 262–265

Vacuum tubes, 26–27, 92–95

Vatz, Mara E., 26, 27, 92–95

Venus Paradise Coloring Set, 12, 252–259, 276–277

Wallpaper, 168–169
Walls, 12, 47–48
Weinberg, Gil, 118–119
"What We Build," 132–166, 248–250
"What We Model," 70–99, 236–240
"What We Play," 102–129, 242–246
"What We Program," 180–215, 262–265
"What We See," 42–51, 220–226
"What We Sense," 54–68, 228–233
"What We Sort," 168–177, 252–259
What's Where in the Apple I, 200
Whitman, Walt, 275, 289n2
Wilder, Thornton, 265
Willow, Diane, 66–68, 279
Winnicott, D. W., 20, 279–280, 286n21, 289nn11, 12
Wizardry (game), 183–185, 199
Wood stove, 84–87
Wright, Frank Lloyd, 242
Wright, Will, 286n27

Yeo, Richard, 285n8
Yoo, Ji, 189–192